Geology of Indiana State Parks

By

Max W. Reams

and

Carol A. Reams

An interpretive guide to the geological side of
Indiana state parks and nature preserves

Disclaimer:

This book does not pretend to be a complete discussion of the subject matter. It is intended for the enrichment of anyone with a general interest in the topics covered. References are included for those who wish to delve more deeply into the subjects. No claim is made about the accuracy of information in this book or about the validity of information in the references cited.

Warning:

State parks are not necessarily safe places to visit. Always take great care for your safety and the safety of others while visiting parks. Be aware of wildlife, steep slopes, and any natural or artificial hazards. The authors or publisher are not responsible for injury or damage caused by your travel to state parks, preserves, or other sites or while you visit these locations, regardless of your understanding of information in this book. Always check weather conditions and heed warnings issued by the state, e.g., ticks, bears, feral hogs, coyotes, snakes, poisonous plants, etc. Always check with staff at the locations for any questions. Heed their recommendations. Indiana parks are usually reached by vehicle, so be aware that many are on rural roads with various hazards. Always carry a printed map for areas without adequate cell phone service to access GPS locations. Some areas permit hunting. Be careful.

Copyright © 2024 by Max W. and Carol A. Reams All rights reserved.

ISBN 9798338592694

Independently Published

In accordance with the U.S. Copyright Act of 1976, the scanning, uploading, and electronic sharing of any part of this book without the permission of the publisher constitute unlawful piracy and theft of the authors' intellectual property. If you would like to use material from the book (other than for review purposes), prior written permission must be obtained by contacting the publisher and authors. Thank you for your support of the authors' rights.

Cover art by Jeremy Spence, Dept. Art & Digital Media, Olivet Nazarene University; B.A., MFA Governors State University.

Cover photos and in text photos © Max and Carol Reams

ACKNOWLEDGMENTS and DEDICATION

Solomon said, "There is nothing new under the sun." (Ecclesiastes 1:9). This book could not have been written without an enormous number of resources. The authors are indebted to a multitude of publications and individuals who provided data used to prepare the text. Consult the References section for specific sources. Especially important is information from publications of the Indiana Geological and Water Survey. Katherine Bulinski and Alan Goldstein provided fossil identification and information on Falls of the Ohio State Park. Photographs are the authors' and are copyrighted. Illustrations based on cited publications are interpreted by the authors. We are grateful to thoughtful and insightful readers, especially Polly Sturgeon of the Indiana State Geological and Water Survey. Their excellent suggestions, careful work, and attention to details have greatly improved the text. All errors are ours. This book is dedicated to our children, grandchildren, and great-grandchildren.

ABOUT THE AUTHORS

Max W. Reams taught geology at Olivet Nazarene University for over five decades. His B.A., B.S., and M.S. are from the University of Kansas, and his Ph.D. is from Washington University (St. Louis). His specializations include the study of caves and sedimentary rocks. In addition to numerous professional publications, his book titles include: *Geology of Illinois State Parks, Waterfalls in Illinois* (with Carol A. Reams), and *Geology of Missouri State Parks* (with Carol A. Reams). Science-based detective novels include: *Oil on My Hands, My Mine or Yours, Diamonds: Friend or Foe, A Fossil By Any Other Name,* and *How Dark is Your Cave?* Additional books include: *Before your Journey* (study guide for premarried couples) and *On the Journey* (study guide for married couples).

Carol A. Reams' bachelor's and master's degrees are from Olivet Nazarene University. She coauthors the books *Waterfalls in Illinois* and *Geology of Missouri State Parks* with Max W. Reams.

Table of Contents

Introduction 1

Safety in the Parks 3

The Big Picture 7

What makes physiographic regions different? 11

Bedrock: The Foundation of Indiana 12

Minerals: The Stuff of Rocks 12

Rocks: Three Kinds 16

Mineral Resources 26

Weathering and Erosion 27

Rock Units: Dividing Them Up 34

Bedrock Structures of Indiana 39

Plate Tectonics: How the Earth Works 41

Earth Structures 42

Fossils in Indiana Rocks 46

Scenery: What To Look For As You Travel 47

Soils and Vegetation: Life Essentials 61

Overview of Indiana Geologic History 63

PHYSIOGRAPHIC REGIONS 78

Overview of Indiana Physiographic Development 79

1. Northern Moraine and Lake Region 83

    1a. Lake Michigan Border 89

       Indiana Dunes State Park and Indiana Dunes National Park 91

       Examples of Lake Michigan Border nature preserves 97

    1b. Valparaiso Morainal Complex 100

       Example nature preserves in Valparaiso Morainal Complex 101

    1c. Kankakee Drainageways 104

Tippecanoe River State Park 108

Potato Creek State Park 111

Example nature preserves in Kankakee Drainageways............. 114

1d. St. Joseph Drainageways 118

Example nature preserves in St. Joseph Drainageways 119

1e. Plymouth Morainal Complex 120

1f. Warsaw Moraines and Drainageways 122

Example nature preserves in Warsaw Moraines and
Drainageways 123

1g. Auburn Morainal Complex 126

Pokagon State Park 128

Chain O' Lakes State Park 133

Example nature preserves in Auburn Morainal Complex 139

2. Maumee Lake Plain Region 144

2a. Maumee Lake Plain 144

Example nature preserve in Maumee Lake Plain 146

3. Central Till Plain Region 147

3a. Bluffton Till Plain 147

Ouabache State Park 149

Example nature preserves in Bluffton Till Plain: 150

3b. Iroquois Till Plain 156

3c. Tipton Till Plain 158

Prophetstown State Park 159

Example nature preserve in Tipton Till Plain 160

3d. New Castle Till Plains and Drainageways 161

Mounds State Park 163

Summit Lake State Park 167

    Whitewater Memorial State Park 172

    White River State Park.. 177

    Fort Harrison State Park 178

    Example nature preserves in New Castle Till Plains and
    Drainageways.. 179

  3e. Central Wabash Valley 182

    Turkey Run State Park 185

    Shades State Park 195

    Example nature preserves in Central Wabash Valley................ 199

4. Southern Hills and Lowlands Region 206

  4a. Wabash Lowland... 208

    Shakamak State Park 212

    Harmonie State Park.. 214

    Example nature preserves in Wabash Lowland......................... 217

  4b. Boonville Hills 218

    Lincoln State Park 220

    Example nature preserves in Boonville Hills............................. 222

  4c. Martinsville Hills 223

    Example nature preserves in Martinsville Hills 225

  4d. Crawford Upland 232

    O'Bannon Woods State Park 237

    Example nature preserves in Crawford Upland........................ 241

  4e. Mitchell Plateau.. 247

    McCormick's Creek State Park 250

    Spring Mill State Park 254

    Example nature preserves in Mitchell Plateau.......................... 258

  4f. Norman Upland 265

Brown County State Park	269
Example nature preserves in Norman Upland	276
4g. Scottsburg Lowland	277
Example nature preserve in Scottsburg Lowland	280
4h. Charlestown Hills	281
Charlestown State Park	283
Falls of the Ohio State Park	286
Example nature preserve in Charleston Hills	292
4i. Muscatatuck Plateau	293
Versailles State Park	295
Clifty Falls State Park	299
Example nature preserves in Muscatatuck Plateau	307
4j. Dearborn Upland	309
Example nature preserve in Dearborn Upland	310
Conclusion	312
Glossary of Terms	313
References	326
Understanding Topographic Maps	333
How Waterfalls Form	334
Index of Parks, Regions, and Preserves	335

Introduction

Indiana has a wide range of beautiful landscapes set aside as state parks. These parks are among the most well maintained and cared for of any state. Their quality is retained with the help of entry fees. Excellent books describe Indiana parks, their geology, history, and recreational value, e.g., Strange, 2018; IGWS, 2018; Higgs, 2016, 2019; Camp and Richardson, 1999.

So, why another book about Indiana state parks? This book is designed for anyone who wants to understand something about how the varied physical features in the parks originated without having to plough through technical treatises or books which cover the geology of the state in varying detail. When you visit a park, you should be able to turn to that park in this compact book and, in a short time, grasp the essence of how this amazing area came to be. You don't need a degree in geology to appreciate the natural processes which produced the beauty you see.

This book follows the general design of *Geology of Illinois State Parks* (Reams, Geology of Illinois state parks, 2013) and *Geology of Missouri State Parks* (Reams & Reams, 2022). Toss this book into your vehicle's storage compartment or drop it in a backpack, so it is available when you visit an Indiana park.

The introductory topics cover essential concepts of geology related to Indiana. A glossary of terms is included. References are available to explore a particular topic or area more thoroughly.

Unlike most books on parks, this one is less oriented toward human history, wildlife and plants, discussion of trails, and recreational opportunities. Those can be found on any park's website or brochure. The emphasis here is on the geologic background and physical history of Indiana parks. Have fun exploring amazing features in these places set aside for citizens to enjoy!

In addition to Indiana state parks, we include geological information on some nature preserves and other natural sites. Be aware that such areas may have similar or greater restrictions as state parks. There is no collecting of anything but trash. Vandalism of any kind is forbidden. Nature preserve trails or lack thereof may make hiking difficult.

Always read about any park or site online before visiting it to be aware of closures, ease of access, dangers, or other issues. For example, some parks or areas may allow hunting in season. This can pose a danger to visitors. Always carry a map or trail guide when visiting any area. GPS may or may not be accessible in some places. A road map may also be needed as a backup to locate some parks or sites in remote regions.

Safety in the Parks

One of the dangerous assumptions often made is, "It's a park, so it must be safe." Nothing could be further from the truth! Scenery is almost by definition hazardous. Steep cliffs, slippery trails, loose rocks, tree roots, poisonous plants, dangerous animals, and a host of other hazards can surprise the unwary visitor. Although nothing is guaranteed, here are a few suggestions which may help as you prepare for your visit to a park or nature preserve.

Take safety precautions, especially those provided by local park staff, websites, and brochures. Wear appropriate clothing, footwear, and hats or caps. Verify your physical ability to follow trails. Stay on marked trails. Pay attention to the "rugged" nature of trails. Some can be hazardous (Figure 1). Pay close attention to park officials or other authentic sources concerning safety and rigor of trails. Many nature preserves have minimal trail maintenance and some have no trails. See websites/brochures/local staff to understand any hazards and avoid dangerous locations.

Never climb waterfalls. Every year, numerous individuals are injured or die as a result of climbing waterfalls in the United States. Don't let the height of a waterfall be a guide to safety. Their slippery nature can be deadly on even small waterfalls (see (Reams & Reams, 2021)).

Although you may not see any of the following wildlife and plants, take necessary precautions, such as traveling in parks with others. Lone hikers are more vulnerable to injury or other problems. Examples of wildlife to avoid include this incomplete list.

Figure 1. Read about trail ruggedness or availability on brochures or websites before hiking. (Portland Arch Nature Preserve)

Rabid animals: seek immediate medical attention if bitten.

Deer are the most dangerous animals in Indiana, primarily because of collisions with vehicles.

Coyotes usually avoid humans, but do not trust them.

Cougars are rare in the Midwest. Recent Illinois capture and roadkill data indicate they are migrating into their natural range and habitat. Avoid them. Report sightings to Indiana Division of Fish and Wildlife.

Bobcats are rarely seen, but always avoid contact with them.

Bears were extinct in Indiana by the 1870s, but have been seen in the state in 2015, 2016, 2018, and 2021. Black bears are making a comeback in the United States, although it is unlikely you will see one in Indiana. If you do, watch from a car or building and do not disturb the animal. Never attempt to feed a bear or any wild animal. Report your sighting to the Indiana Division of Fish and Wildlife.

Feral hogs are primarily found in southern Indiana where they cause agricultural and ecological damage. Avoid them since they can be dangerous to humans.

Venomous snakes which can occur in Indiana parks are primarily copperhead, timber rattlesnake, cottonmouth/water moccasin, and eastern massasauga rattlesnake. Watch your step and avoid them. If bitten, seek immediate medical attention.

Poisonous plants include poison hemlock, cow parsnip, wild parsnip, poison ivy, poison sumac, and stinging nettle. Seek medical attention as appropriate.

Ticks are most prominent in certain habitats during the peak season of early April into July. See websites on how to avoid and treat ticks, including medical concerns.

Chiggers are primarily found in southern Indiana. See websites to avoid and how to treat if bitten.

Spiders such as brown recluse and black widow are the only venomous varieties known in Indiana. Seek medical attention immediately if bitten by a spider.

Insects of various kinds may bite or sting. Take precautions and seek medical treatment as needed.

For any of the above and others not listed, avoid contact with questionable organisms. If affected, respond appropriately to avoid medical issues.

There is no collecting of anything in the parks or preserves except trash. Treat other visitors with respect. Never deface natural or human-made features. Leave no evidence of your visit except on a sign-in book, website comment, or by taking a photo. Enjoy God's amazing creations set aside by the State of Indiana and its residents for your enjoyment, relaxation, education, recreation, and meditation.

The Big Picture

To appreciate Indiana from a natural physical history viewpoint, let's begin by looking at the whole state. When you view the landscape as you travel, what you see, besides trees, buildings, and ranger stations, is *topography*, or "lay of the land." In other words, you see how the elevation changes from place to place. These are not random variations. They are the direct result of physical processes acting on or in the Earth.

Some parts of Indiana are fairly flat, while others consist of rugged landscapes. On a smaller scale, landscape is a composite of an almost infinite number of details and processes spread out over geologic time. The outcome results in an amazing array of beautiful and wonderful features. The variety of colors and shapes can stimulate our senses and produce an awe for the Creator's hand.

We are wired to classify things. Geographers traditionally divide the land surface into areas that have similar natural features. These are called *physiographic provinces*. Provinces are in turn subdivided into smaller units called *physiographic sections* and these in turn are broken into smaller subdivisions.

Traditionally, Indiana was partitioned by geographers into two provinces. The area which experienced coverage by the *Ice Age* (Pleistocene) *glaciers* is part of the Central Lowlands Province, a huge area in the middle of North America. This province in Indiana was divided into two sections, the Great Lakes Section and the Till Plain Section. The unglaciated area in southern Indiana was designated part of the Interior Low Plateaus, a province that extends south into Kentucky and beyond (Hunt, 1967).

Modern geographic research takes an expanded approach to dividing the physiographic regions of Indiana (Gray, Physiographic Divisions of Indiana, Indiana Geological Survey Special Report 61, 2000). We will follow Gray's method in discussing the state's *physiography* (Figure 2).

Each physiographic area has unique physical attributes. Indiana state parks are distributed in many of these areas. The geologic history of each area affects the appearance of local landscapes. For this reason, we discuss the parks, nature preserves, and other natural areas within each physiographic unit since they have common origins. Not all physiographic areas are blessed with a state park. Nature preserves and other natural areas will be mentioned to provide background to further emphasize each region's characteristics.

Indiana State Parks by Physiographic Region

1. **Northern Moraine and Lake Region**
   a. **Lake Michigan Border**
      Indiana Dunes State Park
   b. **Valparaiso Morainal Complex**
   c. **Kankakee Drainageways**
      Tippecanoe River State Park
      Potato Creek State Park
   d. **St. Joseph Drainageways**
   e. **Plymouth Morainal Complex**
   f. **Warsaw Moraines and Drainageways**
   g. **Auburn Morainal Complex**
      Pokagon State Park
      Chain O' Lakes State Park
2. **Maumee Lake Plain Region**
3. **Central Till Plain Region**
   a. **Bluffton Till Plain**
      Ouabache State Park

b. **Iroquois Till Plain**
  c. **Tipton Till Plain**
     Prophetstown State Park
  d. **New Castle Till Plains and Drainageways**
     Mounds State Park
     Summit Lake State Park
     Whitewater Memorial State Park
     White River State Park
     Fort Harrison State Park
  e. **Central Wabash Valley**
     Turkey Run State Park
     Shades State Park
4. **Southern Hills and Lowlands Region**
  a. **Wabash Lowland**
     Shakamak State Park
     Harmonie State Park
  b. **Boonville Hills**
     Lincoln State Park
  c. **Martinsville Hills**
  d. **Crawford Upland**
     O'Bannon Woods State Park
  e. **Mitchell Plateau**
     McCormick's Creek State Park
     Spring Mill State Park
  f. **Norman Upland**
     Brown County State Park
  g. **Scottsburg Lowland**
  h. **Charlestown Hills**
     Charlestown State Park
     Falls of the Ohio State Park
  i. **Muscatatuck Plateau**
     Versailles State Park
     Clifty Falls State Park
  j. **Dearborn Upland**

Geology of Indiana State Parks

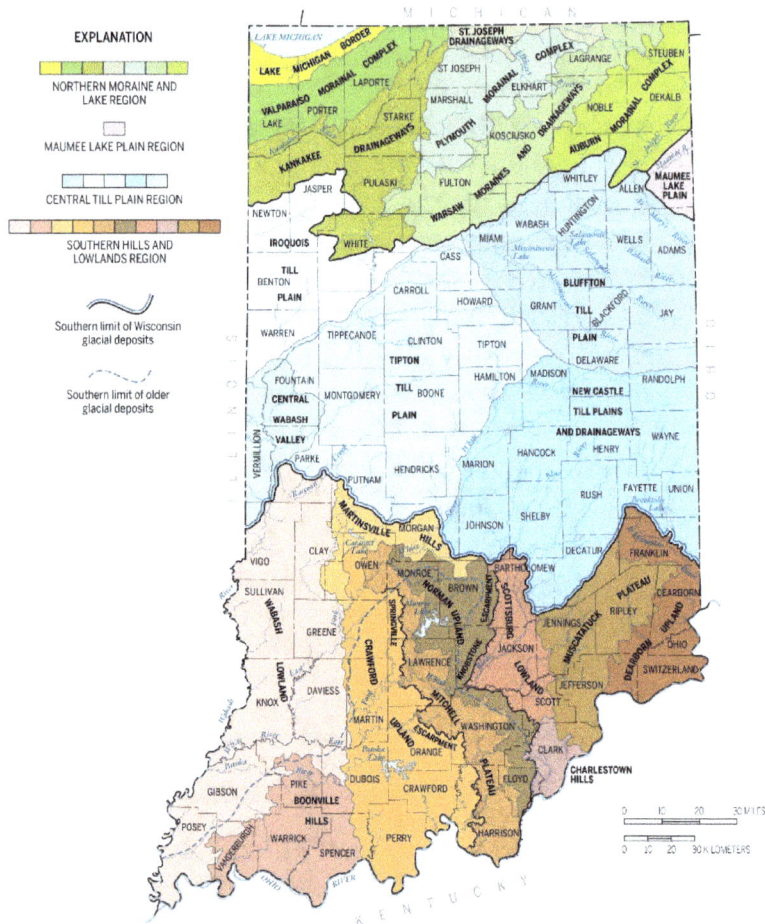

Figure 2 Map of Indiana showing physiographic divisions: Indiana Geological Survey Miscellaneous Map 69, scale 1:1,624,615, Gray, H. H., 2001. © Indiana Geological and Water Survey, Indiana University, Bloomington.

What makes physiographic regions different?

Each physiographic unit's landscape is a special combination of *tectonic activity* (uplift or downwarping of the land, faulting, or fracturing of rocks), *bedrock* type (solid rock masses that underlie the area), *erosion* (by glaciers, streams, wind, waves, etc.), loose *sediments* (deposited by glaciers, streams, wind, waves, etc.), *soils* (formed by weathering of local sediments or bedrock), and *climate* (average temperature and precipitation). It is the combination of these factors through geologic time that results in the landscape and vegetation of a region.

There is a striking boundary between northern and southern Indiana shown in Figure 2. A sharp demarcation is drawn at the southernmost advance of the most recent glaciers (called the Wisconsinan glaciation). Glacial deposits south of this line mainly date from previous glaciations (Illinoian and pre-Illinoian glaciations). The names of these glacial events are taken from states where glacial evidence was first discovered for each. North of the Wisconsinan glacial boundary is a relatively flat surface. South of this boundary there is a region where glaciers have never touched.

Bedrock: The Foundation of Indiana

If you dig down any place in Indiana, you eventually strike solid rock. This is the foundation on which surface sediments and soils rest. Cities and buildings are built on bedrock, sediment, or soil.

What type of bedrock you might hit as you dig is the product of natural processes in the geologic past. These rocks can range from deposits laid down in ancient shallow seas to molten magma that cooled long ago. All rocks are made of *minerals*, which are natural chemical compounds. Examples include *quartz, feldspar, calcite, dolomite*, etc.

Minerals: The Stuff of Rocks

Quartz

Quartz (SiO_2) is not the most common mineral on our planet. However, it is one of the most durable. Quartz displays strong scratch hardness. This means that quartz can scratch most other minerals. Rub a piece of quartz on a nail and it leaves a scratch mark on the metal. The durability of quartz makes it a common residue on the Earth's surface (*sand, silt*).

The color of quartz is variable. Most grains or crystals are clear or white, while others are red, purple, or yellow. Few minerals are as resistant to the weathering effects of water and gases of the atmosphere as quartz. As a result, most sand-and silt-size grains in *sandstone* and *shale* rocks in Indiana are made of quartz. Quartz is originally formed in igneous or metamorphic rocks (Figure 3). After weathering of these rocks at the Earth's surface, the main solid residue is quartz. Other minerals in igneous and metamorphic rocks decompose to clay or dissolve in water.

Many Indiana limestones and similar rocks contain masses or layers of *chert*, which is a very fine-grained deposit of quartz. Flint is a variety of chert. Chert in Indiana is precipitated directly by groundwater or by replacing minerals like calcite or dolomite in sedimentary rocks. Chert comes in many colors and shapes. Chert breaks when pressure is applied to form a *conchoidal fracture* that produces a very sharp edge where two fractures meet. Chert was the preferred material used by skilled Native Americans for making tools of many useful shapes.

Feldspar

Feldspars (various chemical formulas) are the most abundant minerals making up the Earth's crust of igneous and metamorphic rocks. Feldspar colors include pink, white, or gray. Feldspar is softer than quartz and slowly decomposes in contact with atmospheric gases and water. It can weather out of igneous and metamorphic rocks as sand. However, most feldspar eventually is weathered to *clay minerals* and quartz. There are no bedrock exposures of igneous or metamorphic rocks in Indiana. Glacial *gravels* and boulders scattered over the state contain some of these rocks with visible feldspar grains (Figure 3).

Calcite

Most animal shells are made of calcite ($CaCO_3$). This mineral can be clear, white, or yellow brown. Calcite is very soft and easily scratched by quartz or feldspar. To distinguish quartz from calcite, scratch the minerals with a nail. Calcite will be scratched but not quartz. The sedimentary rock *limestone* consists of ground-up shells of calcite. Calcite in limestone can easily dissolve in rainwater or groundwater. Once groundwater is saturated with the dissolved mineral, the water may redeposit calcite in open spaces in bedrock. Groundwater deposits calcite in air-filled *caves* to form *speleothems* of many shapes (*stalactites, stalagmites, rimstone dams, draperies, cave pearls*, etc.). *Aragonite* is a mineral similar to calcite. It makes up shells of snails, clams, corals, etc. Aragonite is also found in some speleothems.

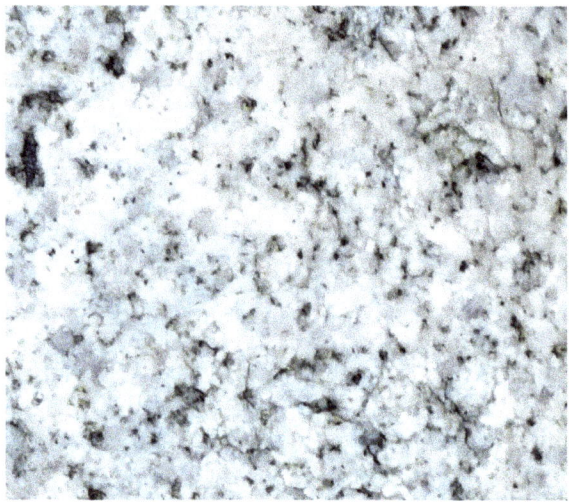

Figure 3. Closeup of a granite boulder brought from Canada by a glacier. Quartz grains are gray. Feldspar grains are white. (Pokagon State Park)

Dolomite

When calcite is exposed to salty groundwater, calcite may change to dolomite, $CaMg(CO_3)_2$, by exchanging half of the element calcium for the element magnesium. Dolomite looks much like calcite. The best way to tell them apart is by a chemical test. Dolomite is easily dissolved in rainwater and groundwater but reprecipitates as calcite, not dolomite. Dolomite is not found in speleothems; instead, calcite is the main mineral in cave speleothems. *Dolostone* is a sedimentary rock made of the mineral dolomite. Dolostones are common in older sedimentary rocks of Indiana.

Clay Minerals

The most common sedimentary rock is *shale*. The main minerals making up this rock are tiny grains of quartz and flakes of clay minerals that are too small to be seen without significant magnification. Clay minerals are produced by weathering most types of rocks. Clay minerals are an important part of soils. They are significant in agriculture due to their reactive flat surfaces that allow the exchange of important elements for use by plants.

Rocks: Three Kinds

Rocks are made of minerals. There are three broad categories of rocks: igneous, metamorphic, and sedimentary.

Igneous Rocks: Cooled Lava or Magma

Rocks that form by cooling of molten material are called *igneous rocks*. These melted masses (*magmas*) form deep in the Earth at high temperatures necessary to melt rocks. Magmas are less dense than solid rock and are buoyant, allowing these liquids to push their way upward. In the cooler part of the Earth's crust, they may solidify. A common rock that cools at depth is *granite*, which is pink, gray, or whitish-colored and made mainly of large crystals of feldspar and quartz (Figure 3). Some magmas are so hot they push completely through the Earth's crust and pour onto the surface as *lava*. If lava has the same chemical composition as granite, it cools quickly to form *rhyolite*. Rhyolites have the same minerals as granite, but the crystals are tiny due to fast cooling.

A second type of magma forms at higher temperatures than granite. It is hotter, heavier, and has a different composition. This magma may push to the surface and form a lava rock called *basalt*. It is black and made of tiny crystals. If this magma cools slowly at depth, it is *gabbro*, a coarse-grained dark-colored rock. A third type of lava has an intermediate composition between basalt and rhyolite. This fine-grained rock is *andesite*.

Igneous bedrock in Indiana is found only by deep drilling through overlying sedimentary rocks. These ancient (Precambrian) rocks include basalt, granite, gabbro, and andesite (Rudman & Rupp, 1993). Igneous rocks can be found at the surface in gravel and as glacial boulders (*erratics*) in Indiana. These have been brought from Canada by glaciers during the Pleistocene (Figure 4).

No igneous bedrock is exposed in Indiana. The closest surface exposures are in the St. Francois Mountains of eastern Missouri where the Earth's crust was uplifted and stream erosion removed younger, overlying rocks.

Figure 4. Erratic boulders of igneous (left) and metamorphic rocks (right with cross-cutting igneous dikes) eroded from Canada by Pleistocene glaciers and deposited in Pokagon State Park.

Metamorphic Rocks: Reheated

Rocks buried deep in the Earth experience high temperatures, pressures, and stresses. Altered igneous or sedimentary rocks display minerals arranged in layers and flattening of minerals that may be contorted. These are *metamorphic rocks*. There are no surface exposures of metamorphic bedrock in Indiana. Deep drilling in Indiana reveals ancient (Precambrian) metamorphic rocks. Examples include marble (metamorphosed limestone or dolostone) (Bieberman & Esarey, 1946) and *gneiss* (contorted layers of minerals) (Rudman & Rupp, 1993). Pleistocene glaciers picked up fragments of metamorphic rock as far north as Canada and transported them to Indiana. They can be found in gravels and as boulders (erratics) in glacial deposits (Figure 4).

Sedimentary Rocks: From Sediments

Rocks exposed to the Earth's atmosphere decompose by physical and chemical weathering by ice, water, air, and organisms into sediment (clay, silt, sand, or gravel) and dissolved minerals. Sediments are transported by wind, water, gravity, or glaciers and are deposited in streams, lakes, *floodplains*, caves, dunes, *deltas*, lagoons, beaches, or the ocean. If these loose sediments are buried in the Earth, they become compacted and groundwater precipitates mineral matter between grains. This produces a solid *sedimentary rock* such as *sandstone*.

Chemical weathering products removed from rocks can be used by animals and plants for nutrition and to build shells, bones, etc. Concentrations of shells buried and solidified form limestone. Chemically modified limestones form dolostones. Thick masses of buried vegetation can be compacted and altered to make *coal*. Other buried organic matter can form *petroleum* and *natural gas*. Layers of sedimentary rocks form *strata*.

Sandstone

Sand is derived from granite and metamorphic rocks as quartz and feldspar grains. Quartz is durable and makes up most sand and silt-size particles in sandstone, siltstone, and shale. Feldspar, which is easily decomposed, is found in sandstones that have not experienced much weathering. Sand can be deposited on river floodplains, deltas, dunes, beaches, and on the shallow sea floor. When sand is buried, groundwater typically precipitates the minerals quartz or calcite in spaces between grains. This cements grains together, forming *sandstone*. Colors of sandstone range from white to gray to tan to brown to red.

Sandstones may show *cross-bedding* (angular layers within a flat stratum) (Figure 5). Cross-beds form where a current of water or wind deposits sand on a sloping surface. Well-cemented sandstone is resistant to weathering and forms ledges on the sides of hills in Indiana (Figure 6). Weakly cemented sandstones usually form gentle slopes. Sandstone may contain fossil shells or plants.

Figure 5. Two sets of cross-bedded layers in horizontal Pennsylvanian Mansfield sandstone. Cross-bedded units form a flattening boundary at their contact. (Portland Arch Nature Preserve)

Sandstones are found in many Paleozoic Era sequences of sedimentary bedrock strata in Indiana, particularly in the western and southwestern portions of the state. Deep drilling revealed an apparent Precambrian sandstone (Rudman & Rupp, 1993).

Conglomerate

Gravel deposits that are cemented with quartz or calcite by groundwater are called *conglomerates*. They are often deposited as layers in ancient stream or beach sediments. Sandstones and conglomerates may occur in alternating layers in Pennsylvanian Period strata.

Figure 6. Pennsylvanian Mansfield sandstone forms a ledge over a weakly resistant shale layer. (Williamsport Falls)

Siltstone

Siltstone is made of extremely fine-grained quartz. The difference between sandstone and siltstone is based on grain size. One sixteenth of a millimeter is the official boundary. For practical purposes, silt is barely visible to most unaided eyes. Well-cemented siltstone and sandstone form a prominent ridge in southern Indiana called the Knobstone Escarpment (Figure 7).

Figure 7. Mississippian siltstone from the Knobstone Escarpment, southern Indiana. Silt grains are barely visible to unaided eyes.

Shale

Mud is very fine-grained weathered material derived from many types of rocks. It consists of tiny grains of quartz silt mixed with clay minerals. The grains are usually so small as to be invisible to an unaided eye. Mud is easily transported by streams and waves and may be deposited in lakes, river floodplains, glacial deposits, deltas, lagoons, tidal flats, or the ocean floor. Buried mud deposits have the water squeezed out by pressure from overlying sediments. This bonds the grains to form shale. Shales can be gray, black, red, pink, tan, or green.

Shales exposed to the atmosphere easily weather back to mud. Shales usually form gentle slopes on hillsides and are covered with vegetation. Shale also forms recessed areas below stronger sandstone, conglomerate, limestone, or dolostone, which form resistant ledges (Figure 8). Shales may contain fossil shells or plants. Shales are found in many Paleozoic Era strata in Indiana.

Figure 8. Thick Pennsylvanian Period gray shale under thin, brown sandstone layers. Shale weathers faster than sandstone and is recessed while sandstones form ledges. (Cagles Mill Lake spillway)

Limestone

Shells of clams, snails, and other invertebrates are either made of calcite or aragonite. Ground-up shells are cemented together to make limestone by groundwater depositing calcite in spaces between shells. Limestones often contain well-preserved fossils. Limestones are more resistant to weathering than shales. However, they are soluble and groundwater dissolves the rock along fractures

Geology of Indiana State Parks

and weak areas (Figure 9). This produces thousands of *sinkholes* (surface depressions) (Figure 10) and caves (passages in bedrock) in Indiana. Limestones are common in Paleozoic Era strata throughout the state.

Figure 9. Devonian limestone with solutional pockets dissolved by groundwater. (Falls of the Ohio State Park)

Dolostone

Calcite in limestones altered by concentrated salt water may be converted to form dolomite. Rocks made of dolomite are called dolostones. Many times in geological literature "dolomite" is used in place of "dolostone" when referring to the rock rather than the mineral. This is confusing, but geologists have gotten used to the terminology. Fossil shells are often not as well-preserved in dolostone as in limestone due to the conversion process of calcite to dolomite. Dolostones are common in older Paleozoic strata (Figure 11).

Coal

Coal is an organic deposit of altered plant remains found in Indiana's Pennsylvanian Period strata. These coals formed in swamps near the shore, often in deltas or lagoons, where stagnant waters allowed vegetation to slowly decompose. Coal is found as part of Pennsylvanian *cyclothems* (rhythmic deposits of alternating sandstone, coal, limestone, and shale). These cyclic deposits were deposited on top of each other as sea levels rose and fell. The changes in sea level were due to glacial cycles of warming and cooling in the southern hemisphere. During sea level rise, swamp deposits were buried under other sediments. As burial progressed, pressure and temperature increased. Vegetation broke down chemically. Gases escaped, as did organic liquids, leaving a solid organic residue and fossil particles. The remaining matter is solid, black coal. Coal exposed at the Earth's surface weathers easily and is rarely visible in roadcuts (Figure 12). Coal has been a major source of non-renewable energy. Indiana is among the top ten states consuming and producing coal. Additional coal for electrical production comes from other states.

Figure 10. A sinkhole south of Spring Mill State Park.

Geology of Indiana State Parks

Figure 11. Dolostone bedrock. Depressions form by uneven dissolving of rock. (Anderson Falls Nature Preserve).

Figure 12. Coal bed (black) on sandstone. Above the coal are shale and sandstone of a cyclothem. (Cagles Mill spillway)

Mineral Resources

Indiana has a rich mineral resource heritage due to the varied environments in the state's geologic history. Besides coal, oil, and natural gas for energy, Indiana is a major producer of sand, gravel, limestone, dolostone, cement, clay, gemstones, gypsum, and peat (U.S. Geological Survey, 2019). One might say that God has blessed Indiana with many commodities to serve human populations.

Weathering and Erosion

Breaking down and removing rocks

The way for rocks to be seen at the Earth's land surface is by uplift of the Earth's crust or lowering of sea level, followed by removal of overlying materials by weathering and erosion. *Weathering* involves the breakdown of rocks and minerals by water, ice, gases in air, organic chemicals, microbes, and plant and animal activity. These processes form sediment, (e.g., clay, silt, sand, gravel) and dissolved mineral matter in water. Erosion is the removal of sediment from an area by glaciers, streams, wind, waves, or gravity.

Loose Sediments in Indiana

Bedrock has a strong influence on Indiana's landscape, mainly in the southern regions. North of the Wisconsinan glacial boundary, glacial *till* covers the bedrock surface and fills ancient river valleys. Other loose sediments include wind-blown deposits of silt (*loess*) blown off floodplain river deposits. Loess is usually thickest on a river's *bluffs* and thins with distance from the source of the dust, the floodplain. Much of central Indiana is covered by broad plains of till and alluvium deposits, while major areas of loess lie east of the Wabash and White Rivers. *Sand dunes* are found on parts of floodplains, along the southern shoreline of Lake Michigan, and in areas of catastrophic flooding where glaciers melted, like the East Fork of the White River near Brownstown. *Talus* piles of loose rock accumulate at the bases of steep cliffs. South of the Wisconsinan glacial boundary, bedrock can be seen or is covered by a thin residue of ancient glacial sediment, weathered rock, loess, or soil. This area has been weathered and eroded for an immensely longer time than the northern glaciated areas. Thus, the southern soils are highly leached and are not usually as agriculturally productive as the rich younger soils to the north.

Glacial Deposits

Climate moves in cycles with cold and warm periods alternating over long periods of time. If the average temperature is low enough, then snow accumulates and builds up from year to year. Pressure from added snow results in snow recrystallizing to ice. Once ice is thick enough, it flows under its own weight, producing a glacier. Today, glaciers are primarily restricted to high mountains and to latitudes near the poles with low average temperatures.

Glaciers move over the Earth's surface and pick up soil and rocks. These are transported to the edge of the glacier where melting and evaporation of ice deposit till, a loose mixture of clay, silt, sand, and gravel (Figure 13). Streams of glacial meltwater carry some sediment and deposit it downstream (Figure 15). Wind blowing over dried stream sediment picks up and deposits silt (loess) as the wind dies down.

Glaciers moved south from Canada during the great Pleistocene Ice Age. The Pleistocene Epoch began about 2.6 million years ago and ended abruptly 11,700 years before the present. Studies suggest the air temperature rose rapidly over a period of only a few decades as the Earth passed from the Pleistocene into the Holocene Epoch (today's warmer time). Early geologists recognized four major southward advances of ice sheets into the Midwest. These were named for states where till exposures were first known (from oldest to youngest): Nebraskan, Kansan, Illinoian, and Wisconsinan.

These ice advances were separated in time by warm *interglacials*. Today, the terminology most frequently used for *glacials* in the Midwest is pre-Illinoian, Illinoian, and Wisconsinan. All of these subdivisions of the Ice Age are represented in Indiana glacial deposits. Pre-Illinoian glaciations have resisted easy subdivision due to erosion removing these ancient tills.

Figure 13. Melting ice front of a modern Alaskan glacier depositing till made of boulders, gravel, sand, silt, and clay.

Stream Deposits

Streams have huge erosive power. They can carry vast amounts of sediment during floods (Figure 14). Streams carrying gravel and running on bedrock can carve *potholes* in rock. The swirling water moves gravel in a circular motion as stones dig a hole (Figure 15).

Figure 14. Glacial meltwater from modern Alaskan glaciers carries clay, silt, sand, and gravel downstream.

Figure 15. Pothole carved by swirling water carrying gravel. (Turkey Run State Park, photo courtesy of M.S. Stillman)

Rivers draining glaciated regions carried large quantities of clay, silt, sand, and gravel through watersheds of the Mississippi, Missouri, and Ohio Rivers. Large *floodplains* (flat surfaces between valley bluffs where streams wander about and periodically flood during high rainfalls) are seen in the valleys of the Ohio, Mississippi, and Wabash Rivers (Figures 16 and 17).

Figure 16. Bluffs of the Ohio River. (Overlook by IN 62)

Figure 17. Wabash River floodplain of alluvium covered with trees. (Harmonie State Park)

Large amounts of sediment are temporarily stored in floodplain valleys. These alluvium deposits are mobilized during floods and carried downstream. They are replaced with sediment from upstream deposits as floodwaters slow down. Flooding was heavy during Ice Age summers and major warming periods. As glacial meltwaters disappeared at the end of the Pleistocene, streams relied on local rainfall. However, today's rivers retain evidence of Pleistocene flooding. Wide valley floors of many Indiana streams, as well as the Mississippi and Ohio Rivers, are vestiges of larger discharges of glacial meltwater (Figure 18). Rivers today are colored by sediment, organisms, and organic matter. Green is associated with *algae*. During floods, the Ohio River ranges from dark brown to green.

Figure 18. Big Blue River valley formed by glacial catastrophic meltwater flooding. The half-mile wide floodplain has a tiny stream masked by trees in the distance. (West of Summit Lake State Park)

Wind and Gravity Deposits

During dry winter months of Wisconsinan and Illinoian glaciations, cold winds blew over floodplains to pick up silt and deposit loess on bluffs of the Ohio River (Figure 19). Glacial floods spread sheets of silt, sand, and gravel over low areas. Winds heaped up sand into dunes (Figure 20). Freezing and thawing of water in bedrock fractures breaks rock into boulders, gravel, and smaller fragments. These rocks fall by gravity and accumulate at the base of cliffs, forming steep rock piles of loose rock (talus) (Figure 21).

Figure 19. Red-orange loess caps yellow Mississippian sandstone in a quarry near the Ohio River in Illinois.

Figure 20. Vegetated sand dune. (Kankakee Sands Nature Preserve)

Figure 21. Talus pile at the base of a cliff where rocks fell as a result of weathering. (Cagles Mill spillway)

Rock Units: Dividing Them Up

Humans are born to classify things. Geologists name unique rock units based on age, location, or characteristics of strata. Quaternary glacial sediments can sometimes be dated by radiometric methods. Bedrock sedimentary units are not easy to measure this way. Instead, their ages are described in relative terms, i.e., which rock unit is older or younger than another rock unit. Examples of methods used to classify rocks by relative age are these.

1. Order of rock strata in terms of stacking. This principle assumes that sediments are deposited by gravity in a regular manner as they settle out of water, wind, or are dropped by melting ice, so that younger rocks are on top of older rocks.

A sedimentary rock unit is the *formation*. Each formation has distinctive characteristics (color, grain size, fossils (Figure 22), chemical composition, etc.) and is named after the location where it was first described, e.g., the Beaver Bend Limestone was named for a conspicuous bend in Beaver Creek near Huron, Indiana. A formation is a three-dimensional rock layer that extends for a limited distance in all directions. Some cover a few square miles while others extend over thousands of square miles.

A related collection of formations is called a *group*. A subdivision of a formation is a *member*. The orderly arrangement of rock units into a vertical summary is called a *Geologic Column* (Figure 23). They are arranged by order of their relative age.

2. Some rocks are time markers, e.g., when a volcano erupts, ash can spread over a wide area. Since ash deposits are igneous, some can also be dated radiometrically and serve as a reference wherever they are found.

3. *Fossils* (preserved evidence of ancient life in rocks) are useful in determining the relative age of strata. Formations in Indiana may be full of fossils, e.g., some Ordovician strata (Figure 22), while others have few fossils.

Figure 22. Ordovician limestone with brachiopod shells and colonies of bryozoans. (Thistlethwaite Falls valley)

These methods, and many more, are used to construct Indiana's geologic column. The state agency responsible for studying rocks of Indiana and assembling the information about Indiana's geologic history is the Indiana Geological and Water Survey. This organization provides public access to geological information, as well as gathering and generating data. See their website for details about the state's geology and publicly available educational resources. The United States Geological Survey, Washington, D.C., is also a major resource on Indiana geology.

As you use this book about Indiana's state parks, refer to the Geologic Column (Figure 23) when the text discusses the ages of rock strata. The Column helps place the park and its particular rocks in the bigger picture of Indiana's geologic history.

Figure 24 is a geologic map which shows the distribution of bedrock of the various geologic periods in Indiana. Glacial deposits are not shown. They cover the northern two-thirds of the state.

We try to avoid excessive technical jargon. We could use the term *Period* (a major division of Geologic Time; *Eras* are sets of Periods) after the name of a rock unit, e.g., "Mississippian Period". To simply things, we tend to use only "Mississippian." Likewise, instead of using Formation after the name of a stratum, only the name of the stratum may be used, e.g., instead of "Edwardsville Formation," we may use only "Edwardsville."

To keep from continually referring to the Geologic Column (Figure 23), the name of the Period may be used before the formation name, e.g., "Silurian Bailey Limestone" instead of the cumbersome "Silurian Period Bailey Limestone".

Sometimes geologists use the formal name of a stratum with a rock name, e.g., "St. Louis Limestone" instead of "St. Louis Formation." Which approach is used often follows historical convention. Don't worry about it. It's just a thing geologists do!

Geology of Indiana State Parks

ERA	PERIOD/SYSTEM		MILLIONS YEARS AGO	PREDOMINANT ROCK TYPES IN INDIANA	PRINCIPAL FOSSIL TYPES IN INDIANA
CENOZOIC	QUATERNARY		2.6	Unconsolidated deposits - glacial till, sand, gravel, silt, marl, clay, and peat deposited during and after continental glaciation	Mastodon, mammoth, peccary, dire wolf, saber-toothed cat, gastropods, pelecypods, plants, and pollen
CENOZOIC	TERTIARY		65.5	Unconsolidated sediment consisting of clay, mud, gravel, sand, and silt	Short-faced bear, peccary, camels, snakes, rodents, fishes, birds, and turtles
MESOZOIC	CRETACEOUS		145.5	None present	None present
MESOZOIC	JURASSIC		199.6	None present	None present
MESOZOIC	TRIASSIC		251	None present	None present
PALEOZOIC	PERMIAN		299	None present	None present
PALEOZOIC	CARBONIFEROUS	PENNSYL-VANIAN	318.1	Shale, sandstone, mudstone, clay, coal, limestone, and conglomerate	Lycopods, *Calamites*, seed ferns, true ferns, *Cordaites*, and amphibians
PALEOZOIC	CARBONIFEROUS	MISSIS-SIPPIAN	359.2	Shale, sandstone, siltstone, limestone, and gypsum	Crinoids, brachiopods, cephalopods, corals, molluscs, trilobites, bryozoans, fishes, arthropods, and foraminifera
PALEOZOIC	DEVONIAN		416	Upper part: carbonaceous shale Lower part: limestone, dolostone, and shale	Corals, brachiopods, cephalopods, trilobites, pelecypods, and bryozoans
PALEOZOIC	SILURIAN		443.7	Dolostone, limestone, siltstone, and shale	Corals, stromatoporoids, bryozoans, brachiopods, trilobites, gastropods, pelecypods, crinoids, and eurypterids
PALEOZOIC	ORDOVICIAN		488.3	Upper part: shale and limestone Lower part: limestone, dolostone, and sandstone*	Cephalopods, trilobites, brachiopods, bryozoans, crinoids, pelecypods, and gastropods
PALEOZOIC	CAMBRIAN		542	Sandstone and dolostone*	Trilobites, brachiopods, and sponges
	PRECAMBRIAN		4,600	Granite, marble, gneiss, and other igneous and metamorphic rock types*	Microbes

* Present only in the subsurface

Figure 23. Geological timeline of Indiana: Indiana Geological Survey, 2011, Indiana Geological and Water Survey website, https://data.igws.indiana.edu/?r=68099. © Indiana Geological and Water Survey, Indiana University, Bloomington.

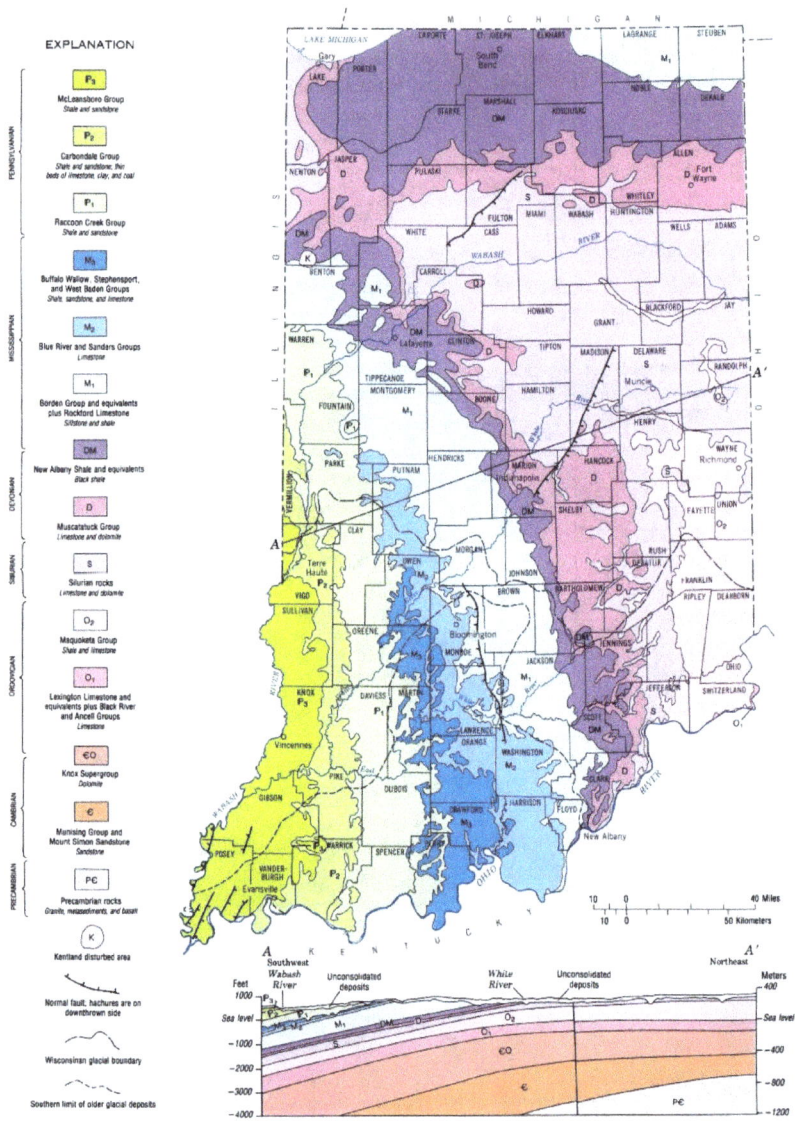

Figure 24. Map of Indiana showing bedrock geology, 1990, Indiana Geological Survey Miscellaneous Map 50, scale 1:1,805,760, © Indiana Geological and Water Survey, Indiana University, Bloomington.

Bedrock Structures of Indiana

The geologic map of Indiana (Figure 24) displays the ages of bedrock that underlie soils, loose surface sediments, and glacial deposits. Sources of this information came from studying bedrock outcrops and thousands of wells drilled for water, minerals, petroleum, and research purposes. Notice the bands on the Geologic Map. They tell the story of how the Earth's crust has been deformed and how erosion has stripped rocks off uplifted areas.

The Geologic Structures Map of Indiana (Figure 25) identifies deformed rock structures shown on the Geologic Map. The northwest-southeast trending structure is the Cincinnati Arch. The smaller feature on the northwest is the Kankakee Arch. These were produced by the bowing up of Earth's crust. The southwest corner structure is the edge of the Illinois Basin, a down-bowed section of Earth's crust. To the north is the Michigan Basin. *Arches* and *domes* are uplifted rock structures and *basins* are down-warped structures. Smaller features are *faults* that appear as heavy black lines. These breaks in the Earth's crust involve shifting of large masses of rocks. How were all these structures formed? To understand the physical causes, let's see how continents are deformed.

Geology of Indiana State Parks

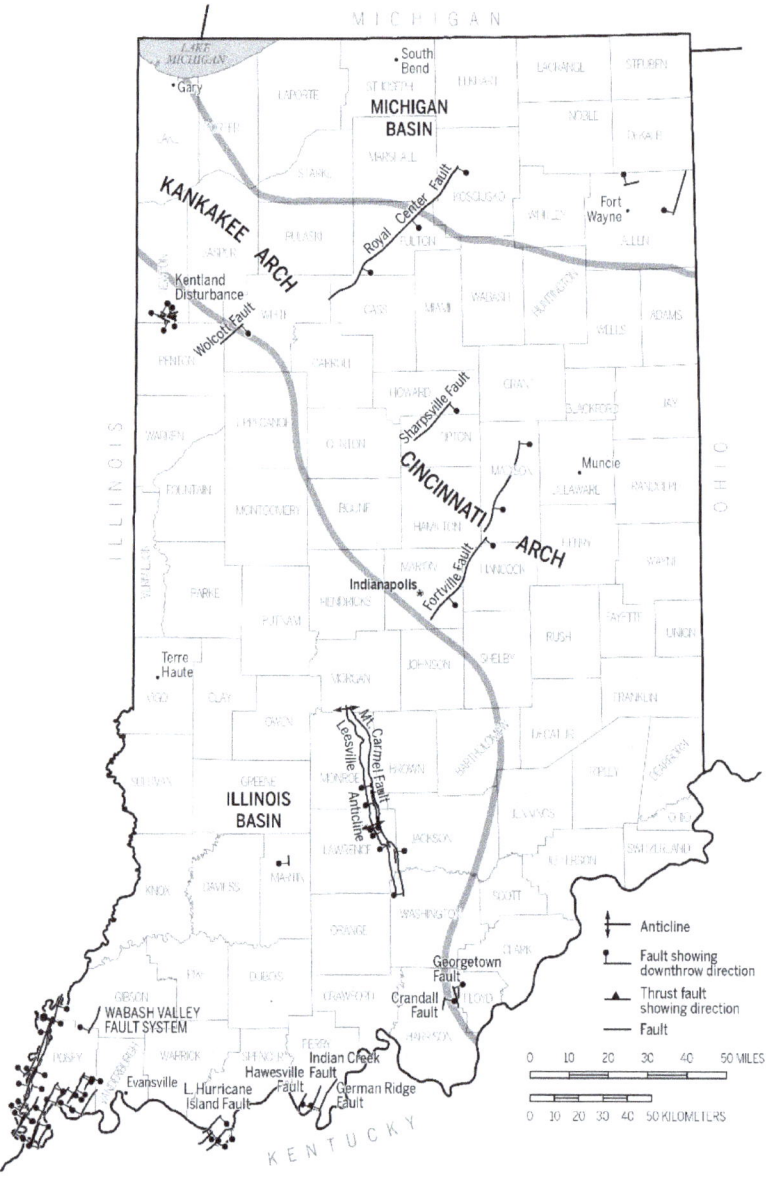

Figure 25. Map of Indiana showing major geologic structures, 2016, Indiana Geological Survey website, https://data.igws.indiana.edu/?r=41847 scale 1:2,070,588. © Indiana Geological and Water Survey, Indiana University, Bloomington.

Plate Tectonics: How the Earth Works

The Earth's interior consists of spherical layers, each with specific types of rocky material. The deepest interior is a solid, iron-rich *inner core* topped by a liquid, iron-rich *outer core*. Overlying the core is the rocky *mantle*. This thick layer is covered by the rocky crust. There are two types of crust:

-*Oceanic crust* is several miles thick.

-*Continental crust* is tens of miles thick.

The crust and upper part of the mantle move together. This set forms a rocky plate. A *plate* can vary in thickness. The average is about 60 miles, with much greater thickness under the continents. There are two kinds of plates:

-Continental plates: made of old, light-weight igneous, metamorphic, and sedimentary rocks.

-Oceanic plates: sedimentary rocks sit on heavy igneous rocks (e.g., basalt) and are covered by a few miles of sea water. The oceanic plate rocks are younger than most continental crust rocks.

Plates move at different speeds but usually have a rate of inches per year. They are mainly driven by movement of hot rock masses at great depths. There are three ways rocky plates move with respect to each other:

1. Plates break and move apart, e.g., North America and Europe move away from each other. The plates diverge most often from an *ocean ridge* where basalt magmas rise to fill the space left as plates spread apart.
2. Plates slide past each other, e.g., the San Andreas Fault of California occurs where the North American Plate moves southeast and the Pacific Plate moves northwest in a jerky manner, producing major earthquakes.

3. Plates move toward each other in three ways:
   a. A continental plate and an ocean plate converge where the heavy ocean plate dives under the lighter weight continental plate. This is *subduction*. Lines of volcanoes (*island arc*) form and erupt as a result. The Cascade Mountains of California, Oregon, and Washington form such a line of volcanoes as the Pacific Plate descends under the North American Plate.
   b. If two ocean plates converge, one ocean plate subducts below the other. A line of island volcanoes forms like the Aleutian Islands off the coast of Alaska.
   c. If two continental plates converge, they are too light-weight to subduct. Instead, they collide and a mountain range is raised, e.g., India collided with Asia, forming the Himalayan Mountains.

In ancient times, the European and African continents slowly collided with North America as did South America. The results of these collisions included the uplift of the Appalachian Mountains on the eastern seaboard and the Ouachita Mountains to the south. Stresses from these collisions were transmitted to the interior of North America. Uplifted arches and down-warped basins in Indiana are the result of these stresses.

Earth Structures

Folds are structures where the rock layers are bowed up or down. In arches, the strata *dip* away from the center line of the fold in two directions. Take a sheet of paper and lay it flat on a table. Push opposite sides toward each other. This forms an arch. A much smaller upfold is called an *anticline*. In basins, strata dip toward the center from all directions. Imagine a nested stack of mixing bowls turned right side up. In *synclines*, strata dip toward the center from two directions. Take a sheet of paper. Hold it in the air by two opposing sides and push toward the center. Now you see a down-folded syncline structure. Most of the folds in Indiana display very slight dips or tilts of the rock layers.

A *fault* is a natural fracture of rocks resulting from stresses that stretch or squeeze the rock until it breaks and the broken blocks of rock slide past each other. Numerous faults are present in Indiana, especially in the southwestern edge of the state. This is near the site of the famous New Madrid earthquakes of 1811-1812. Other major earthquakes affecting the southwest area occurred in 1895 and 1909. Additional earthquake areas are scattered over the state. Faults also occur where meteors strike the Earth (Figure 26).

Figure 26. Fault surface. Rocks moved along the fault to bring two different rock layers in contact. (Kentland Impact Disturbance)

Joints are the most common structural features in bedrock. A joint is like a fault, but the separated blocks do not slide past each other;

instead they separate slightly (Figure 27). Joints often form by expansion of brittle rocks as uplift and erosion removes overlying layers. As a result, most rocks are jointed. These joints are usually seen best when bedrock is exposed to the atmosphere and are enlarged by chemical and physical weathering.

Figure 27. Vertical joint in sandstone and shale. The joint face displays a smooth surface in the sandstone. If this was a fault, the face might have scratches and the rock layers would be offset from one side to the other. Widening of the joint is due to weathering. (Brown County State Park)

Impact structures occur when an extraterrestrial body impacts Earth, e.g., Kentland Disturbance fault in Figure 26. Tremendous pressure formed a cone-shaped structure in bedrock (Figure 28).

Figure 28. Impact feature in bedrock. (Kentland).

Fossils in Indiana Rocks

Fossils are preserved evidence of ancient life. If you are interested in identifying fossils, check out resources on the Indiana Geological and Water Survey website. No collecting of any rocks, fossils, minerals, plants, animals, etc. is allowed in Indiana state parks. The oldest rock with fossils at the surface in Indiana is the Ordovician Trenton Limestone in the southeast corner of the state. Much older rocks with fossils occur below the surface but can be reached only by drilling. The youngest bedrock exposed in Indiana is the Pennsylvanian Mattoon sandstone in the southwest corner.

Animal fossils (*trilobites, brachiopods, bryozoans, crinoids,* clams, *cephalopods,* snails, shark teeth, trace fossils, etc.) are found in Indiana's Paleozoic limestones, dolostones, shales, and sandstones. Plant fossils are mainly preserved in Pennsylvanian shales and sandstones. There are no rocks from the Permian Period and Mesozoic Era found in Indiana. The only Tertiary Period fossils (Pliocene Epoch) come from a sinkhole in Grant County. Extinct Pleistocene Epoch mammals found in Indiana include giant short-faced bear, helmeted muskox, dire wolf, giant beaver, stag moose, giant armadillo, Pleistocene horse, *Mastodon, Mammoth,* Jefferson's giant ground sloth, giant jaguar, saber-tooth cat, and meadow jumping mouse (data from FAUNMAP). See Reams (2022) for details about Pleistocene fossils found in Indiana.

Scenery: What To Look For As You Travel

Physical "scenery" consists of visible land features called *landforms*. These come in a wide variety of shapes and sizes depending on the geologic history of a region, the rock types, and climate, both past and present. There are many landforms in Indiana.

Plains: These are flat land surfaces, e.g., the Iroquois Till Plains in central Indiana (Figure 29). This does not mean there are no hills. Stream erosion accentuates elevation differences. The composition of glacial deposits in Indiana plains has a strong affect in determining *relief* (differences in elevation within an area).

Figure 29. Iroquois Till Plain. A *kame* in the distance marks one of the few departures from relatively flat topography. (US 41 south of US 52 junction)

Geology of Indiana State Parks

Hills and Valleys: Hills are local highs with distinct sloping sides. Hills are formed by a variety of processes. In southern Indiana, most hills are the result of stream erosion, which carves valleys (Figures 30 and 31).

Figure 30. Rolling hills of Brown County State Park.

Figure 31. Valley carved in sandstone of the Pennsylvanian Mansfield Formation. (Turkey Run State Park)

Most northern Indiana hills are due to deposition by glaciers, meltwater streams, or windblown sand. Given enough time, regardless of their origins, hills tend to disappear. Natural processes do the work. Streams erode bluffs. Gullies cut into hillsides. Weathering and gravity-driven landslides remove soil and loose surface debris. The result of these processes is to wear away hillsides. *Sand dunes* are depositional hills formed by wind blowing loose sand into piles.

Moraines: When a glacial *lobe* of ice flows, the glacier erodes sediment of all sizes and soil by incorporating material into the ice and by pushing sediment forward. When the climate warms and less snow is added to a glacier, the ice front stops and begins to melt away. How long a glacier pauses before the ice front melts away determines the height and mass of the ridge of sediment deposited. Eventually, the glacier is said to "retreat." This is a misnomer since glaciers have no reverse gear! The front simply recedes as melting occurs faster than snow is added to the glacier.

The included sediment is dumped onto the land surface by a variety of methods (Figure 32). An ice-contact mixture of clay, silt, sand, gravel, and boulders dropped in front of a glacier is *till*. If the sediment accumulates in a curved ridge outlining the curved shape of the ice front, this is a *moraine*. If this ridge represents the maximum extent of the glacier's forward movement, this is a *terminal moraine*. Due to the complexity of climate history, this is difficult to determine, so the term most used for curved till ridges is *end moraine*. If till was deposited continually as the ice front receded, the uneven surface laid down is called *ground moraine*. Either way, moraines are made of till, i.e., direct deposition of sediment from the ice front.

If a stream of water flows on top of a glacier, sand and gravel may deposit in a fracture in the ice. When ice melts away, sediment is let down onto the land. This forms a mound called a *kame*. Streams flowing in tunnels under glaciers carry sand and gravel. These sediments clog tunnels. This leaves a meandering ridge once the ice front melts away. These ridges are *eskers*. Meltwater streams flowing from the ice front carry vast amounts of clay, silt, sand, and gravel. These sediments are deposited as a sheet or may fill valleys as *outwash* (*alluvium*). Blocks of ice break off the glacial front and may be buried by alluvium or till. Eventually, the blocks melt and depressions are left in fields of glacial deposits. This depression is a *kettle*. Kettles may fill with water (*kettle lake*). Eventually, the lake may fill with sediment and vegetation. Organic matter accumulating in the lake becomes *peat*.

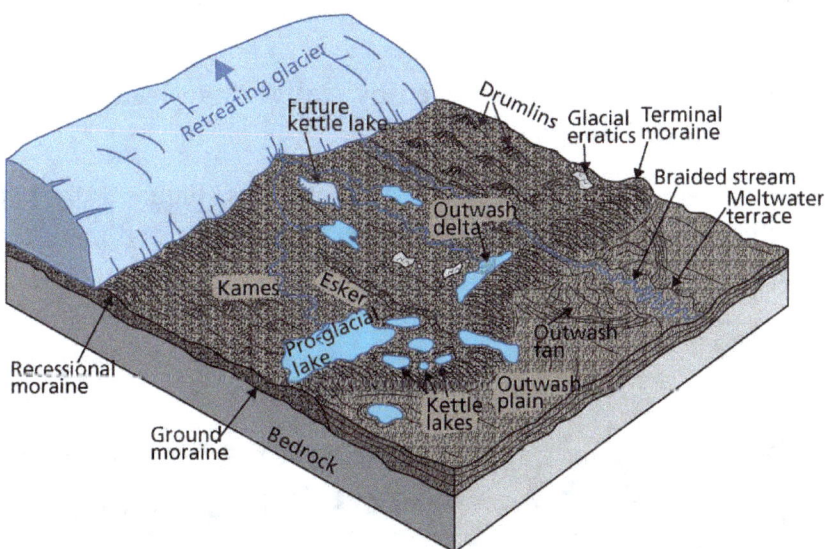

Figure 32. Graphic illustrating glacial depositional landforms.
https://garystockbridge617.getarchive.net/amp/media/glacial-features-and-deposits-6865ee

Mounds: These are semi-rounded hills produced by Native Americans who built significant *mounds* for religious, festival, societal, or burial purposes (Figure 33).

Figure 33. Mounds constructed by Native Americans. (Angel Mounds State Historical Site)

Bluffs: Steep edges of river valleys form bluffs (Figure 34). They are the present limit of lateral erosion by a stream and mark the boundary of a floodplain with the surrounding hills.

Figure 34. Bluffs along the Ohio River. (Across from O'Bannon Woods State Park)

Floodplains: These flat surfaces on valley bottoms are periodically flooded by their meandering streams. The surface is marked by shallow ponds and wandering streams that move back and forth from one valley bluff to the other. Floodplain sediments make fertile soils for farming (Figure 35).

Figure 35. Wabash River floodplain and bluffs. (North of New Harmony).

Meanders: Streams that wander on floodplains form curves called *meanders* (Figures 36-38). Water on the outside of a meander curve speeds up and erodes a steep stream bank called a *cut bank*. Water on the inside of a meander slows down, depositing sediment to form a sand or gravel bar called a *point bar*. With time, these processes accentuate meanders into elaborate S-shaped curves. When meanders of the same river touch, that section of the meander loop can be cut off, forming an *oxbow lake* (Figure 38).

Figure 36. Meander on Lost River south of Spring Mill State Park. Erosion on the outside of the meander causes slumping of the stream bank. Sediment is deposited on the inside of the meander.

Figure 37. Fourteenmile Creek near Charlestown State Park. Sediment is deposited on the inside of a meander (point bar) and erosion takes place on the outside of the curve (cut bank). Two sandy point bars are visible on different meander curves.

Geology of Indiana State Parks

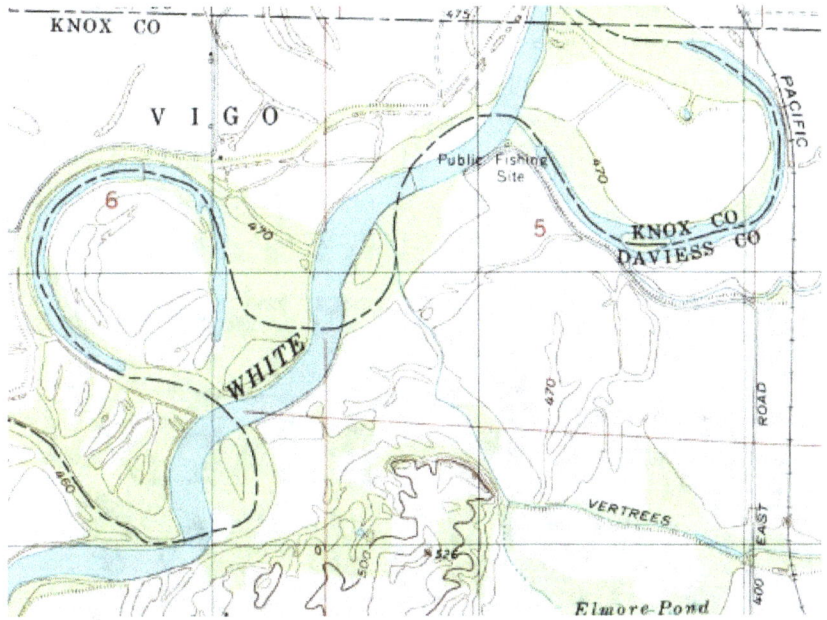

Figure 38. Two oxbow lakes formed by meander cutoffs on the White River. (U.S. Geological Survey Lyons Quadrangle, 1959)

Karst, sinkholes, and caves: Surface depressions that result from collapse of cave roofs or enlargement of vertical fractures in limestone or dolostone are *sinkholes*. They funnel water underground to *caves*. Southern Indiana has over 3,000 caves (Frushour, 2012) (Figure 39) and innumerable sinkholes (Figure 40) developed in limestones and dolostones. Landscapes with sinkholes and caves are collectively referred to as *karst*.

Figure 39. Cave in limestone. (Spring Mill State Park)

Figure 40. Sinkhole developed on Mitchell Plain limestones. (South of Spring Hill State Park)

Geology of Indiana State Parks

Natural bridges: Where surface erosion cuts through bedrock in such a manner as to yield a bridge of bedrock with openings to the surface on both sides, this is a *natural bridge*. This can also form where streams or sinkholes have opened a stretch of a cave, so both ends of the cutoff cave are open to the surface. Other examples involve two streams eroding from opposite sides into a ridge (Figure 41). Examples include Portland Arch, Arrowhead Arch, Ravine Arch, and Wolf Natural Bridges in Indiana. Don't confuse these with the geological structure called an arch.

Figure 41. Arch formed by intersection of two streams through a peninsula of sandstone in the Pennsylvanian Mansfield sandstone. (Portland Arch Natural Area)

Speleothems (chemical cave deposits): Soil is rich in carbon dioxide produced by microbes and decaying organic matter. Rainwater sinks into soil, dissolves carbon dioxide, and moves through fractures in limestone or dolostone. This acidic groundwater dissolves calcite and dolomite minerals. As water seeps into a cave, carbon dioxide is released to cave air, which makes the water saturated with calcite, so the mineral is precipitated as a *speleothem*.

Speleothems can take many shapes and forms. *Stalactites* hang from cave ceilings where water enters from fractures. Stalactites often begin as *soda straws* (elongate hollow tubes). These become plugged and develop a cone-shape by water running down the outside of the tube. *Stalagmites* result from dripping water building up a mound on cave floors. *Columns* form where stalactites and stalagmites join into a single unit. *Draperies* are growths with lateral development from a cave ceiling or shelf. *Rimstone dams* are built at the edge of pools of water. *Cave pearls* form where calcite precipitates from water dripping into a shallow pool onto sand grains. *Helictites* are random growth-branching speleothems that precipitate from slow seepage water. The list of possible speleothem shapes is almost endless!

Most Indiana caves have silt, clay, sand, or gravel sediment washed into them from sinkholes or by back-flooding from surface streams. Each layer or thin *lamination* is evidence of a flood event.

Springs: Groundwater from a cave or other opening pours onto Earth's surface as flowing water (Figure 42). There is no magic source of *spring* water. It comes from local rainfall sinking along joints or direct seepage from surface streams (Figure 43). Water not sinking into bedrock runs off as surface streams.

Figure 42. Panorama of Orangeville Rise, part of the Lost River groundwater drainage network that emerges as a spring. (Orangeville)

Figure 43. Lost River, south of Spring Mill State Park, loses its surface flow as water sinks into gravel and sinkholes.

Waterfalls: Waterfalls occur where geologic events or characteristics of rocks create a sharp slope on a stream bed so water is in free-fall (Figure 44).

Figure 44. Anderson Falls. (Anderson Falls State Nature Preserve)

Soils and Vegetation: Life Essentials

Weathering of bedrock and surface sediments is a complex process involving atmospheric gases (mainly oxygen and carbon dioxide), rainwater, and organisms (especially soil microbes). The result is soil, a mixture of geological material, organic matter, water, and gases. Most plants gain their nutrients, aside from water and carbon dioxide, from chemicals released during weathering. Plants grow better on certain soil types derived from specific bedrock or sediments (Figures 45 and 46).

Figure 45. Woodland on sand ridge and marsh in distance. Vegetation zones are highly controlled by soil/sediment and water. (Ivanhoe South Nature Preserve)

Figure 46. Marsh in kettle lake and forest on glacial sediment. Contrasting water sources, drainage, and soil/sediment control vegetation. (Spicer Lake Nature Preserve)

Vegetation in Indiana is native, exotic, or agricultural. The most productive agricultural soils are in the Till Plains and the Great Lakes Sections. Soils derived from young glacial till and loess tend to yield better crops. Older glacial tills and loess would be second best. Unglaciated areas of southern Indiana without loess caps lack fresh input of nutrients and tend to be less fertile. Clays are important sources of nutrients. Young glacial soils tend to be richer in clay minerals. Native vegetation is presently restricted to restored sites or to areas so secluded that a plow has not touched them. These include state parks and other areas set aside to preserve the way Indiana used to look. These sanctuaries allow wildlife and native plants to thrive. Exotic species not native to Indiana and not used for agriculture have been introduced for various purposes or by accident. These varieties may be difficult to control and can replace native species if allowed to reproduce without restrictions. State park and nature site personnel try to keep native plants healthy by eliminating exotic species.

Overview of Indiana Geologic History

Indiana state parks are widely distributed through various physiographic regions. What you see in the parks reflects Indiana's geologic history.

Glaciers advanced into Indiana several times during the Pleistocene Epoch. The most recent maximum incursion of ice is marked on the Physiographic Map (Figure 2). This Wisconsinan southern boundary is an irregular line running roughly east-west. The exact location of the boundary is sometimes difficult to pinpoint because erosion has removed some the glacial deposits at the edge. South of this line there is evidence of pre-Wisconsinan glaciation (Illinoian and pre-Illinoian). Boulding (2022) suggests there have been five pre-Illinoian glaciations. Exactly how many of these ancient glacial advances are present is marred by erosion and a blanket of windblown *loess* that obscures the evidence in southern Indiana.

North of the Wisconsinan boundary, Indiana's bedrock surface is largely obscured by glacial till, meltwater sediments, and loess. If glaciers had not invaded Indiana, the landscape would look quite different. The preglacial landscape buried by glacial sediments can be deduced by drilling wells through the Pleistocene sediments down to bedrock. Pre-glacial valleys over 300 feet deep are invisible at the surface because they are buried by glacial till. Bedrock exposures are rare but do occur where modern streams cut through glacial sediments.

Most bedrock outcrops are concentrated south of the Wisconsinan boundary, especially in unglaciated areas.

Speak to the earth, and it will teach you, Job 12:8 NIV

Geologic history is preserved in the rocks of the Earth. By learning what the rocks have to teach us, we can appreciate the amazing geologic history of Indiana. We'll begin with the earliest known events and work our way to the present. Resources used in the following include Camp and Richardson (1999), IGWS (2018), Gray (2000), and a host of publications from the Indiana Geological and Water Survey, the United States Geological Survey, and other professional publications.

Precambrian Era Earth History

Precambrian rocks are the oldest rocks underlying Indiana. Anywhere you drill, your bit will eventually slow down as it grinds into hard, *basement* igneous or metamorphic rocks of the Precambrian. No Precambrian rocks are exposed at the surface in Indiana. Our knowledge of these rocks is based on a few deeply drilled wells. Only a handful of Indiana's buried Precambrian rocks have been dated by radiometric means. Their ages are less than 1.5 billion years. Are they the Earth's oldest rocks? Not at all. Rocks in Canada are almost 4.3 billion years old. The oldest minerals in Australia are about 4.4 billion years in age. Moon rocks are a bit older. Indiana's Precambrian rocks are similar to those in Missouri's St. Francois Mountains (Reams & Reams, 2022). These rocks are part of the Eastern Granite-Rhyolite Province (Figures 47-50).

There is a huge time gap of a half billion years between cooling of Precambrian igneous rocks and deposition of Paleozoic sedimentary rocks found by drilling in Indiana. The break between Precambrian and Paleozoic Era rocks is a continent-wide erosion surface (*unconformity*). Cambrian Mt. Simon Sandstone overlies Precambrian rocks. This first Paleozoic deposit was laid down by shallow seas invading Indiana during sea level rise. A sandstone below the Mt. Simon may be Cambrian or Precambrian.

Geology of Indiana State Parks

Figure 47. Precambrian Granite. (St. Francois Mountains, MO)

Figure 48. Precambrian rhyolite lava flow. (Highway 72 roadcut west of Fredericktown, MO)

Figure 49. Precambrian basalt (black) dike intrudes older rhyolite. (U.S.67 roadcut south of Fredericktown, MO)

What happened in Indiana during this missing time gap of 500 million years? North America was part of a supercontinent called *Rodinia* (supercontinents are clusters of several continental masses). Indiana must have been above sea level most of this time and experienced a long period of erosion. The landscape probably looked like knobs in the St. Francois Mountains of Missouri. Younger Paleozoic sedimentary rocks that once covered the ancient erosion surface have been removed by erosion (Figure 50).

The supercontinent rifted apart as the rocky plate cracked and broke up. Northwest "Europe" separated from "North America" and an ocean formed between the continental masses.

Geology of Indiana State Parks

Figure 50. Panorama of a Precambrian knob with exposed igneous rocks. This approximates the buried Precambrian landscape in Indiana minus vegetation. (Johnsons Shut-Ins State Park, MO)

Toward the end of the Precambrian, extensive glaciation likely lowered sea level worldwide, exposing an eroded, desolate landscape with residual hills (*knobs*). No land vegetation was present. Plants did not appear until much later in the Paleozoic.

As the glacial period closed, meltwater from ice sheets poured into the oceans, slowly raising sea level and eventually flooding Indiana with seawater. The advancing waters of the Cambrian seas splashed onto the shoreline, reworking clay, silt, sand, and gravel derived from weathered Precambrian rocks. Sand on beaches would become the Mt. Simon Sandstone. In the deepening waters, shells of invertebrates were deposited over the hardening sandstone. Sediments buried the Precambrian landscape as sea level rose.

The following discussion relies on the Generalized Stratigraphic Column of Indiana Bedrock (Thompson, Sowder, & Johnson, 2015) and other sources. Ages assigned to the geologic periods in millions of years ago are from Geological Society of America Geologic Timescale accessible at the following link:
- https://www.geosociety.org/documents/gsa/timescale/timescl.pdf

Paleozoic Era History

Cambrian Period (541-485 million years)

Shallow seas of the Cambrian began to rise about 541 million years ago. Ocean waves lapped onto the edge of ancient "North America" as rising seas began their march onto the continent (other names are used to describe early versions of North America). The long period of weathering and erosion that affected Precambrian igneous and metamorphic rocks provided sediment for the waves to work into sheets of sandstone. Almost everywhere in North America, Cambrian sandstones were deposited on Precambrian igneous and metamorphic rocks. The Mt. Simon and other sandstones stacked on top of each other with a thickness of several hundred to a few thousand feet. Imagine the vast amount of sand generated by weathering of granite and other rocks to provide this sediment spread over much of the continent!

As the shoreline shifted inland, beach and shallow water sands were deposited farther onto the old land surface. Clear, shallow water offshore allowed a proliferation of invertebrate shelled animals to thrive. Shells accumulated in great abundance on the sea floor to be buried by later sediments and solidified into limestone.

When conditions were right, salty waters altered calcite to dolomite, converting many limestones into dolostones. This happened in the Upper Cambrian and continued in the Ordovician, Silurian, Devonian, and Mississippian Periods. Not all limestones were converted to dolostones. Details of the conversion process from limestone to dolostone by magnesium-rich fluids are still unclear. Alternating with these limy rocks were shale and sandstone. Think of the incredible number of brachiopods, bryozoans, clams, trilobites, cephalopods, etc., needed to account for thousands of vertical feet of dolostones and limestones spread over the entire state of Indiana! God is lavish in creativity! No Cambrian strata are visible in Indiana except by deep drilling.

Ordovician Period (485-444 million years)

Mud and sand were washed off the Appalachian mountains and spread west to be interspersed between Ordovician limestone and dolostone beds in Indiana. These mountains rose by plate tectonic activity as moving plates closed an ancient ocean and an arc of volcanic islands collided with ancient "North America". This was the beginning of a series of plate tectonic events when continental fragments began to collide with ancient "North America" in the process of forming a supercontinent. At least four erosional breaks (unconformities) occurred during the accumulation of Ordovician rocks. These sea level drops due to tectonic uplift exposed Indiana to stream erosion. None of these breaks are visible in rocks exposed at the Earth's surface today. They are seen in drill cores and in states where older Ordovician rocks are exposed. The Cincinnati Arch began to rise as the crust was down-warped to form the Illinois and Michigan Basins. The St. Peter Sandstone is a rock unit laid down on an unconformity. Although only exposed in a *quarry* near Kentland, exposures are seen in Missouri, Illinois, and Wisconsin (Figure 51). This rock is over 99 percent pure quartz and is the product of sea level rise as beach and wind-blown sand covered an eroded land surface. The oldest rocks exposed in Indiana are Ordovician limestones and shales in the southeast corner. They were also deposited on an unconformity. Fossils are common in Ordovician strata (Figure 52).

An erosion surface separates Ordovician and Silurian rocks. At the end of Ordovician and start of Silurian times, mud washed off the eastern mountains. Fluctuations in sea level shifted where shale or limestone were deposited. The see-saw motion of sea level generated alternating shale and limestone before a major glaciation lowered sea level worldwide perhaps 500 feet! This exposed the area to erosion. Later melting of the ice raised sea level and shallow seas again covered much of North America. This formed the unconformity separating Ordovician and Silurian strata.

Figure 51. Ordovician St. Peter Sandstone is 99 percent pure quartz. (Pacific, MO)

Figure 52. Ordovician marine fossils horn corals (*Grewingkia sp.*), bryozoans, and brachiopods (*Vinlandostrophia sp.*, formerly *Platystrophia sp.*). (Whitewater Formation, Thistlethwaite Falls) (Identifications by Katherine Bulinski and Alan Goldstein)

Silurian Period (444-419 million years)

Dolostone and limestone dominate Silurian strata, with some shale and chert (Figure 53). Fossils and coral reefs can be seen where Silurian rocks are exposed on the Cincinnati and Kankakee Arches if not covered by glacial deposits (Figure 54). Because of their porosity, some reefs allowed petroleum to accumulate. Drilling into buried reefs has yielded oil or gas production.

Figure 53. Thin-bedded, cherty Liston Creek Limestone Member of Silurian Wabash Formation. (Seven Pillars of the Mississinewa)

Figure 54. Quarry exposure of Silurian reef on Kankakee Arch. Tilted strata are not due to tectonic activity. These sediments were washed off the reef by waves. (Kankakee County, IL)

Devonian Period (419-359 million years)

Stresses from mountain building events in the Appalachian region resulted in uplift and erosion in Indiana (Cincinnati Arch) as evidenced by the unconformity separating Silurian and Devonian rocks. Until the end of the Devonian, limestones and dolostones were the main rocks deposited. Some contain chert. A notable stratum is the New Albany Shale, a black, organic-rich shale that has puzzled geologists ever since it was discovered. Underground it is a *source rock* for petroleum and natural gas due to its high organic content. The shale spans Upper Devonian and Lower Mississippian time. This rock records stagnant ocean conditions that allowed organic matter to accumulate. Similar conditions occur in the Black Sea today, except this modern sea is a deep-water environment. The New Albany contains tree logs and other land-based plant fossils indicating a close proximity to the shoreline and shallow-water conditions.

Mississippian Period (359-323 million years)

After the New Albany Shale accumulated, Mississippian seas cleared of stagnant conditions to allow a fossiliferous normal marine environment. In this luxuriant setting, invertebrates thrived, their shells and exoskeletons forming limestones (Figure 55). Marine limestones (some with chert) dominated until sediments from the east were deposited to form sandstones and shales in a delta environment like the Mississippi Delta. Deltaic sandy deposits with shale became more the theme, while marine deposits of limestone diminished. A major lowering of sea level broke the depositional pattern before Pennsylvanian rocks eventually took over.

Figure 55. Mississippian Ste. Genevieve Limestone. (Upper Cataract Falls)

Pennsylvanian Period (323-299 million years)

Plate collisions to the south and east raised the Ouachita Mountains and Appalachian Mountains. Sediments from these highlands spread into "North America's" interior. Pennsylvanian was called "Upper Carboniferous" (versus "Lower Carboniferous" for the Mississippian) or "Coal Measures" when first described in the British Isles. For coal to form, the marine environment gave way to a terrestrial collection of deltas, floodplains, swamps, and surface streams. Coal accumulated in stagnant swamps associated with deltas throughout the Pennsylvanian. Shale, sandstone, coal, and infrequent limestone characterized this time period (Figure 56) Geologists recognized repetitious sequences of rocks due to rise and fall of sea level on a cyclic basis. The repeated sequences became known as "cyclothems." Shorelines shifted back and forth to generate these rocks. Glaciers that formed and receded in the southern hemisphere are considered the cause of cyclothem sequences as sea level moved up and down.

Figure 56. Thick-bedded Pennsylvanian sandstone capped by a coal/shale/sandstone sequence cyclothem. (Cagles Mill Spillway)

Permian Period (299-252 million years)

There are no rocks of Permian age known in Indiana. During the Paleozoic, the continental masses that had been added to ancient "North America" formed the supercontinent *Pangaea*. This broke up following the Paleozoic to eventually yield continents somewhat similar to what we have today. This summary ignores a huge amount of geologic history, but details are not necessary to understand Indiana's general geologic background.

Mesozoic Era Earth History

Triassic Period (252-201 million years)
Jurassic Period (201-145 million years)
Cretaceous Period (145-66 million years)

There are no rocks of these ages in Indiana. The state was likely above sea level. Erosion must have been the order of the day.

Cenozoic Era Earth History

Tertiary Period (also called Paleogene and Neogene Periods) (66-2.58 million years)

The Tertiary is traditionally divided into Epochs (oldest to youngest: Paleocene, Eocene, Oligocene, Miocene, and Pliocene). The only known Tertiary in Indiana is a sinkhole deposit of Pliocene fossils, perhaps five million years old.

Why are some geologic periods not represented in Indiana's rocks? Tectonic uplift, erosion, and lack of environments to deposit rocks of these ages are the main reasons for an incomplete Geologic Column.

Quaternary Period (2.58 million years to the present)

Pleistocene Epoch (2.58 million to 11,700 years ago)

This is the Ice Age when continent-size glaciers formed, expanded, and melted away, only to repeat the performance many times. The glacial stages are pre-Illinoian, Illinoian, and Wisconsinan . Boulding, 2022, says there may be five pre-Illinoian glaciations. Glaciations south of the Wisconsinan glacial boundary in Indiana are spotty and covered by loess. This contributes to the difficulty of identifying these ancient glacial events.

Tills deposited by glaciers are identified by their particular characteristics and their overlapping positions with respect to other tills. Soils developed on tills during interglacials (the times between glaciations). Although the exact age of the earliest glacial deposits in Indiana is uncertain, Boulding (2022) suggests about one million years ago, whereas Sturgeon, Loope and Russel (2017) use 700,000 years as the beginning of Indiana's glacial history. The Indiana Geological and Water Survey lists these dates (Surficial Geology, IGWS website):
-Wisconsin glaciation: 10,000 to 79,000 years
-Sangamon interglacial: 79,000 to 132,000 years
-Illinoian glaciation: 132,000 to 300,000 years
Loope et al. (2018) date the maximum extent of Wisconsinan glaciation in central Indiana at 24,000 years ago.

A map of Indiana's rivers shows that most drain to the south, southwest, or west. This is no coincidence. Glaciers moved in arc-shaped lobes into Indiana from the north and east. Meltwater from each glacial lobe flowed away from the glacial front, establishing the current drainage. These main waterways became tributaries of the Ohio or Mississippi Rivers.

Lakes formed in front of glaciers as a result of moraines blocking stream flow or ice blocks melting to form depressions (*kettle lakes*).

Silt, clay, and organic matter accumulated in shallow lakes and became sources of peat and "muck." Sometimes the moraines gave way or were overtopped and catastrophic floods swept through areas downstream. These flooding waters transported huge sheets of sand over large areas. Flooding events also scoured river channels to considerable depths. Two significant series of floods are the Maumee Flood/Torrent (about 17,000 years) and Kankakee Flood/Torrent (about 19,000 years). After the sands dried, wind moved loose sands into fields of dunes. Around the edge of Lake Michigan, waves washed and sorted glacial till and other sediments to create beaches. Wind blowing over these beaches heaped up very large sand dunes.

Paleoindians apparently crossed the *Land Bridge* between Siberia and North America in the Late Pleistocene and migrated south to populate Alaska, Canada, the northern lower 48 United States (including Indiana), Central America, and South America. Holocene Native North Americans and Native Central and South Americans are descendants of those early travelers.

Holocene Epoch (11,700 years to the present)

Climate changed dramatically over only a few decades beginning 11,700 years ago, based on measurements of glacial ice cores in Greenland. That marked the end of the Pleistocene Ice Ages. The great continental ice sheets which covered parts of Alaska, Canada, and the lower 48 United States began to rapidly melt away over the next several thousand years. Sediments deposited since the Ice Age ended are Holocene. These include stream, lake, beach, sinkhole, and dune deposits. Most of the present landscape of Indiana is experiencing stream erosion. Native Americans were the only humans present in Indiana for thousands of years. Some Native Americans built huge mounds. French explorers entered Indiana in the 1670s.

PHYSIOGRAPHIC REGIONS
State Parks and Nature Preserves

Collecting anything, except trash, is not allowed in state parks.

To illustrate some locales and their physical features, photos and topographic maps are used. For assistance see Understanding Topographic Maps in the back of this book.

This section relies on these references, Gray (2000; 2001), Camp and Richardson (1999), Strange (2018), Higgs (2016; 2019), IGWS (2018), Frushour (2012), Malott (1922), websites of state parks and preserves (state-owned and others), many Indiana Geological and Water Survey maps and publications, United States Geological Survey topographic maps, and professional publications.

Authors subdivide natural areas in slightly different ways, depending on the emphasis. Since this book is about geology, physiographic subdivisions are based on Gray (2001). Physiographic descriptions rely on Gray (2000) and Malott (1922). All Indiana state parks are discussed, and some nature preserves are described to illustrate features found in the regions.

If you visit an Indiana State Park or nature preserve, use of a GPS is recommended, although computer-chosen routes may take you over rougher roads than you prefer. Cellphone reception may not be adequate in some remote areas. A detailed highway map of Indiana is highly recommended as a backup.

Overview of Indiana Physiographic Development

Seawater entered Indiana during Cambrian time. The Middle Run Formation sandstone in the southeast subsurface corner of the state is the only evidence of possible Precambrian sedimentary rocks in Indiana. The earliest Cambrian sandstone in Indiana lacks fossils but may be roughly equivalent to the Lamotte Sandstone in Missouri which has Middle Cambrian fossils. Sea level rose and fell multiple times from Cambrian into the Pennsylvanian. Uplift finally drove the seas away and erosion became universal over Indiana after deposition of the Mattoon Formation (the youngest Pennsylvanian rock in Indiana).

Then followed a 300 million-year gap during which streams eroded through uplifted rocks of the Cincinnati Arch. This exposed rocks as old as Ordovician in part of the arch. No strata in Indiana remain that are younger than late Pennsylvanian (except a Pliocene sinkhole deposit). This time-gap from Late Pennsylvanian to Pleistocene time produced an eroded Indiana land surface over which the glaciers advanced from their birthplaces in Canada. The oldest pre-Illinoian tills may be one million years ago or less.

Where did all the sediments eroded from the land go? The short answer is the same as today, they headed for the oceans. The long answer is more complicated. Since the Earth is a relatively closed system, eroded material was transported toward the ocean where it would be deposited in a variety of environments along the way. There is much more to this story which would take too many pages to recount.

Glacial conditions began in Canada about 2.58 million years ago. Glacial deposits in northwest Missouri have been dated at 2.4 million years, which is early pre-Illinoian (Rovey & Balco, 2010; 2011). Malott (1922) recognized Illinoian deposits but was uncertain about earlier glaciations in Indiana. Boulding (2022) presents evidence for five glacial events that occurred prior to the Illinoian. These pre-Illinoian glacial incursions stretched well into southern Indiana. Boulding also lists an additional early glacial advance that affected only the northern part of Indiana. Boulding notes the existence of a pre-Pleistocene stream deposit in southern Indiana. The dates he assigns to pre-Illinoian tills range from 425,000 to 1 million years. The older preglacial sediment ages he suggests are 1 to 2.6 million years. Pre-Illinoian tills of southern Indiana tend to be patchy, which makes them difficult to assign to specific glacial cycles. Windblown loess blankets many areas, covering tills of various ages, which further complicates their study.

Paleoclimate studies place the Last Glacial Maximum about 29,000-19,000 years ago. The coldest peak was about 20,000 years when average temperatures were about 46°F (today's average is 59°F). This may not sound like much difference, but Earth's climate is a sensitive system that changes dramatically with small numbers. The recognized end of the Pleistocene recorded in Greenland ice is 11,700 years when temperatures increased dramatically (over 7°F).

The land surface that was present when the Pleistocene Epoch arrived with its continent-size glaciers is approximated by Figure 57 (Gray, 2000). Especially striking is the now-buried ancient river valley in the central part of the state. This gorge later filled with glacial sediments and is likely part of the ancestral Teays River system. This massive river, comparable with the modern Ohio River, drained ancient sections of North Carolina, Virginia, West Virginia, Ohio, Indiana and Illinois. It joined the Ancestral Mississippi River in Illinois, which carried water to the Gulf of Mexico (see map of the ancient river system).

https://en.wikipedia.org/wiki/Teays_River

This river system is now buried under hundreds of feet of glacial deposits. Glacial tills so thoroughly cover the Pre-Pleistocene landscape in the northern two-thirds of Indiana that you would never guess a valley 300 feet deep exists below your feet! The thick glacial deposits filling the ancient river valley provide large amounts of groundwater for some Indiana public and private wells. Camp and Richardson (1999) note a controversy suggesting the Teays River itself may have drained north and never reached Indiana.

Loope et al. (2018) dated the Last Glacial Maximum in central Indiana at the southernmost Wisconsinan glacial boundary about 24,000 years ago. This was followed by melting back for a few thousand years before readvancing about 22,000 years. Then the glacier melted back again, leaving tills, stream deposits, and lake sediments. How fast did Indiana Wisconsinan ice move? Loope et al. (2018), suggest 40 meters per year when the ice advanced and moved south near the end of glaciation.

The southern limit of Wisconsinan ice marks a dramatic contrast between the lands to the north and south. Wisconsinan sediments are not as badly eroded as their older cousins to the south. Wisconsinan time saw various glacial lobes come and go in Indiana as warm and cold events alternated.

In contrast, the southern section of Indiana has suffered significant erosion between pre-Wisconsinan glacial cycles. No wonder the north and south areas in Indiana are so drastically different! As you visit the many subdivisions of Indiana physiography, enjoy this diversity of landscapes.

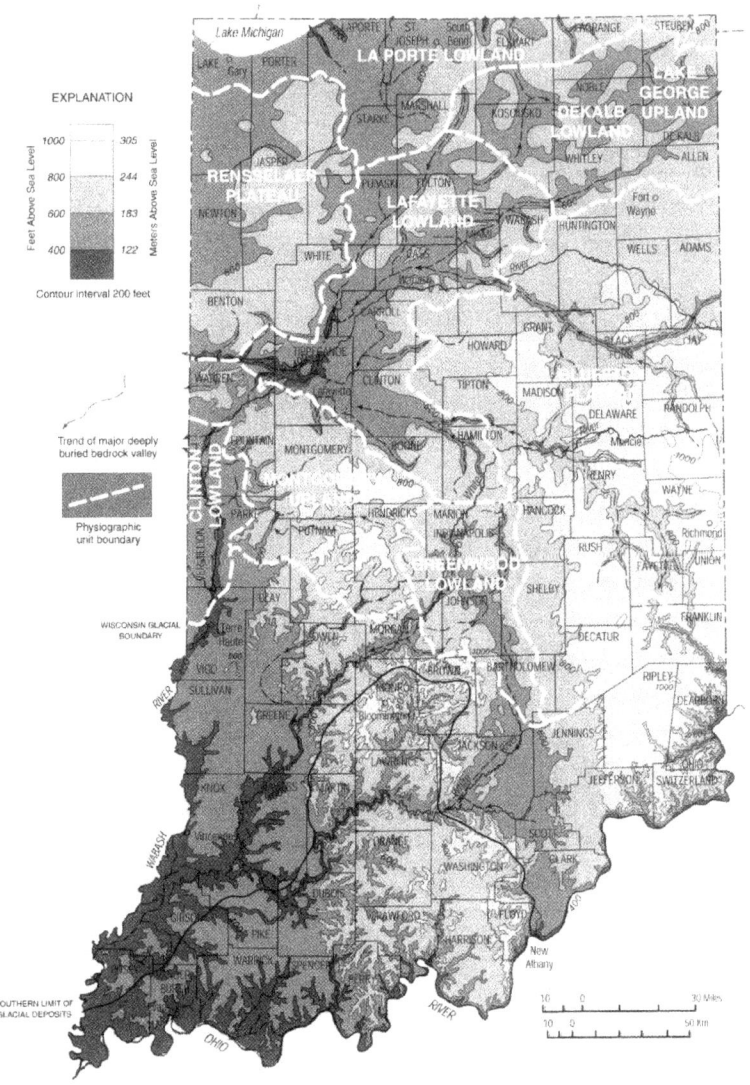

Figure 57. Map of Indiana showing bedrock topography (from Indiana Geological Survey, 1983) and topographic divisions of the buried bedrock surface north of the Wisconsin Glacial boundary, Gray, H. H., 2000, Physiographic divisions of Indiana: Indiana Geological Survey Special Report 61, p. 13, fig. 2, doi: 10.5967/h9a0-bt26 © Indiana Geological and Water Survey, Indiana University, Bloomington.

1. Northern Moraine and Lake Region

This area is dominated by Wisconsinan moraines but includes lake plain and glacial stream drainage deposits. Almost all of Indiana's natural lakes are within its boundaries. Glacial deposits are relatively thick. As a result, bedrock exposures are lacking.

Glaciers erode and deposit sediment. This is clearly displayed throughout northern Indiana. Till in the various moraine landforms is mute testimony to the direct deposition of a mixture of sediment picked up by ice lobes. Canada was the source of snow accumulation needed to generate the glaciers. These moving masses of ice are remarkable erosion agents. Gravel, sand, silt, and clay are gathered from the source areas and more is picked up along the hundreds of miles of movement. These sediments eventually melt out of the ice front to form ridges (moraines) that outline the glacier that laid them down. The named glacial lobes in Indiana came from three different sources: the Lake Michigan, Huron-Erie (a later phase is called the Erie Lobe) and Saginaw (Gray, 2001). Each lobe had a complex history. Once a lobe reached its furthest extent, the ice front paused as a result of climate change. As the source of snow diminished, melting dominated and the ice front began to melt away or "retreat." Left behind was the terminal moraine. If retreat was fairly steady, till was laid down as ground moraine. A pause in retreat allowed a recessional moraine ridge to form. Climate moves in cycles. Additional snow might result in a glacial front moving forward. Warming reversed the movement of the ice front. As a result, deciding if a moraine is terminal or recessional isn't necessarily straightforward. The picture is fuzzy since a glacial readvance may destroy previously deposited tills or

Geology of Indiana State Parks

the ice may override these deposits. Unraveling all this is the task of glacial geologists. Often, glacial ridges are referred to as end moraines or simply moraines.

The Lake Michigan Lobe flowed south from the Lake Michigan Basin and affected the western border of Indiana. The Saginaw Lobe flowed south from Michigan and impacted far northern Indiana. The Huron-Erie Lobe and later Erie Lobe came from the east and affected mainly eastern and central Indiana (Figures 58 and 59). When lobes came in contact with each other, one lobe's deposits would override another lobe's already laid down sediments or interact in complex ways. The landforms from such interactions are different from those involving one lobe.

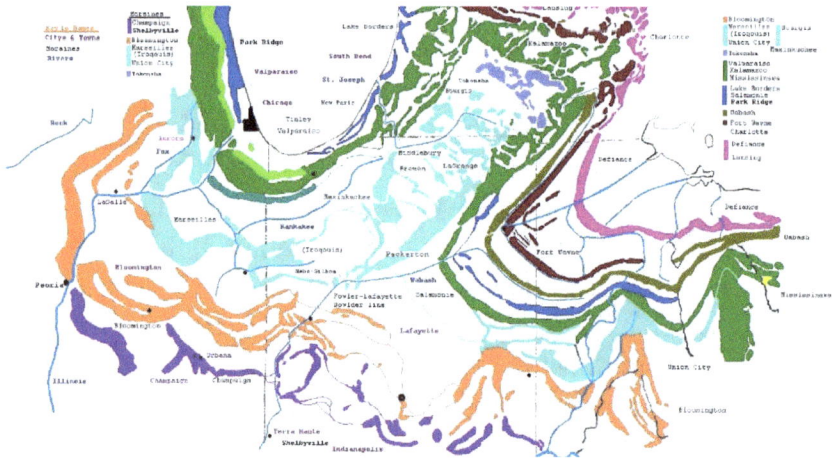

Figure 58. Moraines associated with the Lake Michigan, Saginaw, and Huron-Erie glacial lobes.
https://images.app.goo.gl/6qVVnrnKYRMf1bR46
Wikipedia Commons: Moraine Composite.jpg by Chris Light. Moraines south of Lake Michigan and southwest of Lake Erie. A composite of three maps (Leverett 1915) (Leverett 1902) (Larsen 1986) and other sources. Colors represent moraines from the same time period of the Wisconsin Glacial epoch.

Figure 59. Gently rolling moraine of clay and silt was deposited by the Lake Michigan Lobe. (Potato Creek State Park)

Low places between arc-shaped moraines filled with water to form lakes. The Lake Michigan Lobe retreated to leave Lake Michigan as its legacy. As lobes melted during warming of a glacial cycle, large chunks of ice broke off and were often partly buried by till and meltwater *outwash* sediments. Once these ice masses melted, they left a space or pock mark on the glacial plain. Many filled with water to form *kettle lakes* (Figure 60). Meltwater carried silt and clay into lakes. Most are shallow and support vegetation. Many have no outlets. Some fill with *marl* (a limy precipitate from water), mud (clay and silt), and peat (partly decomposed organic remains). Peat and marl have been mined for commercial purposes.

Figure 60. Big Finster (kettle) Lake. (Chain O' Lakes State Park)

Some meltwater flowed in tunnels under glacial ice. If a tunnel filled with sand and gravel, this formed a cast of the tunnel's shape. Once the glacier melted, this meandering ridge of sediment was left on the Earth's surface as an esker (Figure 61). Some eskers outside protected areas are mined for gravel and sand-building materials.

Figure 61. East-west trending esker between Lonidaw Lake and Charles West Lake in Pokagon State Park. Panoramic north-south profile photo where Trail #6 crosses the esker's crest on the east side. (U.S. Geological Survey Angola West quadrangle, (2019))

Where meltwater flowed on top of a glacier, it picked up sediment melting out of the ice and deposited it in crevasses or low spots on the ice surface. Once the ice fully melted, the pile of sediment left behind was let down onto the land surface. The resulting hill is a kame (Figure 62). Glacial erratics are boulders transported from distances as far as Canada.

A water body like Lake Michigan presented a large water surface to the winds which created waves. Waves broke on the shores where till and outwash sediment were deposited. Beaches formed as waves sorted sediment by grain size with larger and heavier sand concentrated at the shoreline (Figure 63). Thin layers of heavy black sand may be found among quartz sand layers. Finer silt and clay were washed into deeper water where they settled out to form lake muds. Sand on the shoreline is blown by the wind to form tall sand dunes (Figure 64). Lake levels rose and fell, creating sets of features, such as beaches, sand dunes, and *bogs* in low spots. Many different sets of these features formed and were partly or wholly destroyed by wave action.

Figure 62. Road bisects a kame near Chain O' Lakes State Park.

Figure 63. Lake Michigan waves move sand along beach. (Indiana Dunes State Park)

Figure 64. Sand dune by Lake Michigan beach. (Indiana Dunes State Park)

1a. Lake Michigan Border

The Lake Michigan Border area is well known for its sand dunes and beaches along the shoreline of Lake Michigan. Glacial moraines are also present. Due to changes in lake level, the type of sediments being deposited at any moment is directly related to the position of the lakeshore. The dynamic nature of lake level changes, combined with wave action and wind erosion and deposition, make this area a changing landscape. Devonian bedrock underlies the glacial and post-glacial features.

The edge of Lake Michigan reflects the shape of the Lake Michigan Glacial Lobe. Moraines of till around the lake illustrate the curvature of ice margins when ice was at a standstill. A glacial front is dynamic and is always shifting to match the requirements placed on it by the demands of ice supply and melting. The Lake Michigan Lobe of Wisconsinan glacial ice was a massive feature. The compressed snow which fell far to the north recrystallized to solid bluish glacial ice. This was the birthplace of the *Laurentide Ice Sheet*. Estimates of its thickness hover around two miles of slowly flowing glacial ice although it was probably thinner near the edges. This huge mass exerted enormous pressure, depressing the Earth's crust. Weight of the ice, plus the tremendous erosive power of moving ice, scooped out the Lake Michigan Basin.

Modern glaciers move at a wide range of speeds. Some move many tens of feet per day while others only inches per year. Kehew et al. (2005), suggest that the Lake Michigan Lobe was a fast-moving ice mass. The Lobe may have reached its maximum southern extent about 28,000 years ago (Curry, Lowell, Wang, & Anderson, 2018), more than halfway down Indiana's western border. Loope et al. (2018) indicate the Lake Erie Lobe reached its southern-most extent in central Indiana 24,000 years ago. Once the Lake Michigan Lobe receded in steps, the huge weight of ice ceased to locally depress the Earth's crust and rebound began. As the lobe retreated, lakes appeared in front of the receding ice front between moraines.

The first evidence of an emerging lake related to the current Lake Michigan was perhaps 18,000 years ago (Indiana Geological and Water Survey (IGWS), 2018). Geologists have given these ancient lakes names, e.g., Glacial Lake Chicago, etc. The Lake Michigan Lobe retreated and advanced many times from 17,000 to 13,500 years ago before exiting the basin scoured out by the ice 12,000 years ago (Indiana Geological and Water Survey (IGWS), 2018). The land began to rise (rebound) due to pressure reduction on the crust. Lake levels have gone up and down at various times, partly controlled by input and exits of water, as well as rebound of the Earth's crust. The ancient lake which built up behind the Valparaiso Moraine Complex drained via the Des Plaines and Illinois Rivers until the ice melted away from the Lake Michigan Basin. Drainage then shifted to the Straits of Mackinac.

Beaches formed at the lake edges. With each change in lake level, beaches and sand dunes formed in concert with the elevation of the wave action. Detailed changes in lake level are recorded by stranded old beaches and dunes. Many earlier formed beaches and dunes were obliterated or modified by the changeableness of Lake Michigan's level. The maximum lake level of Lake Michigan occurred about 4,500 years ago. A rapid drop in lake level exposed large stretches of sand. Wind blew this sand into dunes that eventually assumed the features seen today.

The Lake Michigan Border is bounded on the south by the Valparaiso Morainal Complex and on the north by Lake Michigan.

Indiana Dunes State Park and Indiana Dunes National Park

Wave action along Lake Michigan's shoreline sorted glacial sediment, washing fine silt and clay into deeper water, where they settled out on the lake bottom as mud. The coarser and less easily transported sand is concentrated along the shoreline (Figure 65).

Figure 65. Waves continually sort and move beach sediment.

Wind moves sand not submerged by water waves (Figure 66) and heaps the sand into dunes.

Figure 66. Wind moves sand in ripples and dunes.

The interaction of waves and wind form the beaches and dunes of Indiana Dunes State Park and Indiana Dunes National Park. The state park of 2,182 acres occupies three miles of the shoreline spared from the ravages of industrialization by those concerned for the loss of public use of the land in 1926. The national lakeshore and its 15 miles of shoreline protection did not appear until 40 years later. Eventually the lakeshore became a national park in 2019 covering 15,349 acres. You can walk the world-class dune trails (Figure 67) and experience the wonder of how they continually change as wind and wave act on the environment.

Figure 67. Sand dunes and beach. Their positions depend on the interaction of wind and waves. (Indiana Dunes State Park)

Dunes shift over time and bury trees. After a boy fell into a hole on Mt. Baldy in the national park, scientists discovered how trees engulfed by moving sand could decay and their residue resist the tendency of sand to fill the hole. Many holes were found, resulting in closure of Mt. Baldy to the public. The dune is moving several feet per year (Figure 68). Wind-driven waves move beach sand parallel to the shoreline (Figure 69). This *longshore drift* transports sand from the east side of the lake toward the south. A breakwater at the Michigan City Harbor traps sand and prevents sediment from moving southward to replenish the parks' beaches.

Figure 68. Advancing sand dunes bury trees. Vegetation decomposes but may leave a hollow tube which creates a hazard for persons walking on the dunes. (Indiana Dunes National Park)

Figure 69. Waves arriving at an angle to beach move sand parallel to beach by longshore drift. Indiana Dunes State Park.

Trucking sand to the sand-starved beaches has tried to alleviate beach erosion. Construction along many Lake Michigan shorelines impede the movement of sand. Certain grasses and brush slow sand dune migration (Figure 70).

Figure 70. Grass and brush slow dune migration. Indiana Dunes State Park)

Figure 71. Great Marsh. Sand dunes block low areas to form wetlands. (Indiana Dunes National Park)

Low spaces between dune ridges form wetlands such as marshes, bogs, and fens (Figures 71 and 72). These pass through various ecological stages. Great Marsh is the largest interdunal wetland in the Lake Michigan watershed. The habitat is critical for breeding and migratory birds. An observation deck is accessible at the north parking area.

Figure 72. Great Marsh. Tall, tree-covered sand dunes in the distance are responsible for blocking drainage to form this huge wetland. (Indiana Dunes National park)

Cowles Bog (National Natural Landmark). This 4,000-year-old wetland is in Indiana Dunes National Park. Henry Chandler Cowles worked out the ecological succession concept for wetlands and popularized the term "ecology." Today, many school children visit the bog to learn about vegetation succession.

Geology of Indiana State Parks

Dunes Nature Preserve (National Natural Landmark): This is the eastern two-thirds of Indiana Dunes State Park and features foredunes, interdunes, and back dunes. **Ancient Pines Nature Area** is a *blowout* which exposes buried trees killed by the dunes. Blowouts occur where vegetation that once held down the dune is removed by human traffic or other forces. The elliptical bowl-shaped depressions are very distinctive (Figure 73).

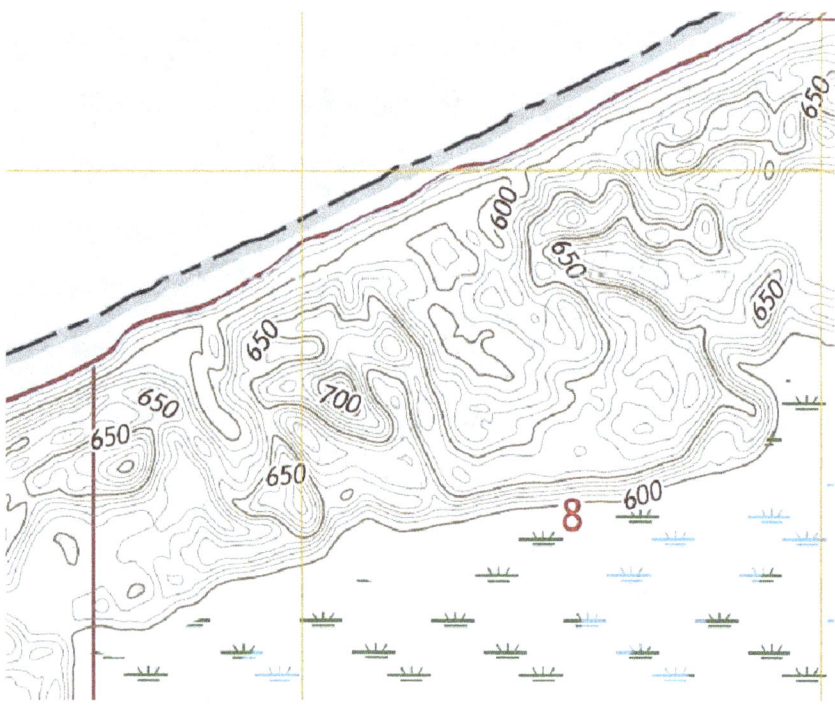

Figure 73. Blowouts (depression contours), dunes (hills), beach and marsh in Dunes Nature Preserve of Indiana Dunes State Park. (U.S. Geological Survey Dune Acres Quadrangle, (2016))

Hoosier Prairie Nature Preserve (National Natural Landmark). This sand prairie was once common in northwest Indiana. Some 350 native plants are known from the dune and depression landscape. It is a unit of the Indiana Dunes National Park.

Examples of Lake Michigan Border nature preserves

Gibson Woods Nature Preserve, Tolleston Strand Plain, Seidner Dune and Swale Nature Preserve, Ivanhoe Dune and Swale Nature Preserve/Ivanhoe South Nature Preserve.
These areas feature rare dune ridges and swales that have almost disappeared from the region.

Ivanhoe South Nature Preserve. Past lake levels each developed beaches and dunes with low marshy areas landward of the dunes. (Figures 74-76).

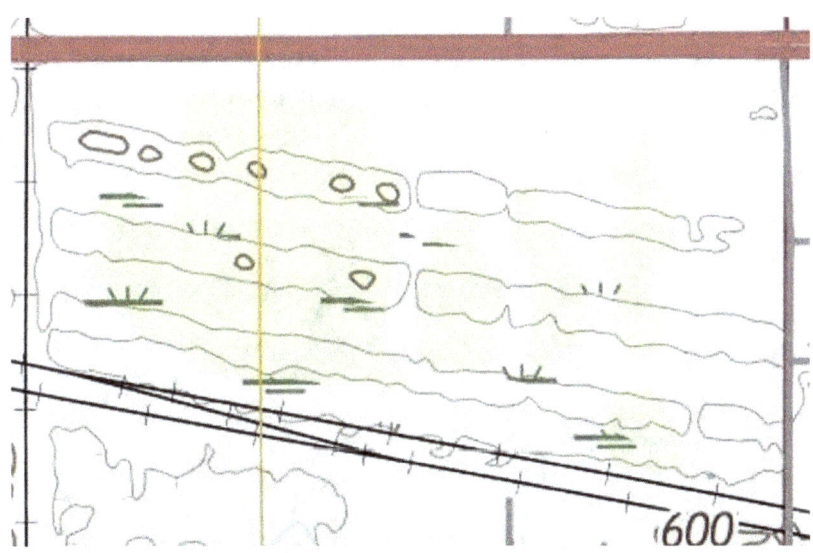

Figure 74. Ivanhoe South Nature Preserve features dune and swale topography north of railroad. Wet, low, swampy areas alternate with linear beach/dune ridges. These reflect changing lake levels and formation of beaches with dunes. A boardwalk crosses a swale between dunes. The boardwalk was not useable May, 2024. (U.S. Geological Survey Highland Quadrangle, (2016))

Geology of Indiana State Parks

Figure 75. Panorama across rare swale and swell topography of wetland, dune/beach ridge, and wetland. Wave and wind processes operating here in the past produced a clean, well-sorted sand like that seen in Indiana Dunes State Park beaches and dunes today. Oak savanna dominates on the ridges.

Figure 76. View of marsh in swale from crest of dune/beach.

Pinhook Bog (National Natural Landmark) This kettle lake formed where a huge piece of ice broke off the glacier and was partially buried in till or outwash sediment. After the ice melted, the depression formed a kettle lake. Lakes gradually fill with moss, plants and sediment. If undisturbed, they can progress from lake to bog to meadow and eventually forest.

John Merle Coulter Nature Preserve This preserve features sand dunes, sand prairie, black oak savanna, sedge meadows, and interdunal wetlands (Figure 77). Over 400 species of native plants are present, as well as many animals.

Figure 77. Sand prairie and black oak savanna are only two of several ecosystems in John Merle Coulter Nature Preserve. All owe their locations to the variety of sand dune-beach/wetland heritage left by retreat of the Lake Michigan glacial lobe in concert with the rise and fall of the lake level.

1b. Valparaiso Morainal Complex

The Lake Michigan Glacial Lobe deposited till as end moraines where the ice front paused. As is common where glaciers are in retreat, a lake formed in front of the ice. During balanced conditions of ice from the northern source of snow and melting at the edge of the lobe, till accumulated as the Valparaiso Moraine. This ridge roughly conforms to the shape of the glacial lobe. During forward movement of the glacial front, previously deposited tills may be overridden or incorporated into the glacial ice or shoved forward. The Tinley Moraine north of the Valparaiso Moraine marks a readvance of ice (Schneider, 1967). The Tinley is smaller and younger than the Valparaiso. The composition of these moraines is nearly identical. They are rich in silt and clay, with a little gravel. These moraines merge to the east and blend to form the Lake Border Moraine. Till ridges are the main topographic features in the Valparaiso Morainal Complex. Rolling landscape is evident on the moraines. Other features include lakes in low spots, alluvial fans of sand, and kettle lakes where ice blocks melted to leave depressions. Lakes tend to fill with mud and marsh vegetation. Meltwater in lakes broke through to carry water into the Kankakee Drainageway to the south. Bedrock below the glacial deposits is Devonian.

The Valparaiso Morainal Complex is bounded on the north by the Lake Michigan Border and on the south by the Kankakee Drainageways (Figure 2).

There are no state parks in Valparaiso Morainal Complex.

Geology of Indiana State Parks

Example nature preserves in Valparaiso Morainal Complex

Moraine Nature Preserve has no established trails but features rolling hills, steep ridges, deep-wooded gorges, potholes, and a small kettle lake.

Spicer Lake Nature Preserve's 320 acres features swamp forest, marsh, and two kettle lakes (Figures 78-80). An isolated block of glacial ice melted and formed the large depression. This wetland has been filled in with vegetation to the point where only two separate lakes remain. There are trails plus a quarter-mile boardwalk with handicap access. The Spicer Lake boardwalk provides a cross-sectional experience across the kettle through ecological successions of swamp forest, shrub zone, and emergent zone by the lake. An impressive list of plant and animal species is available online.

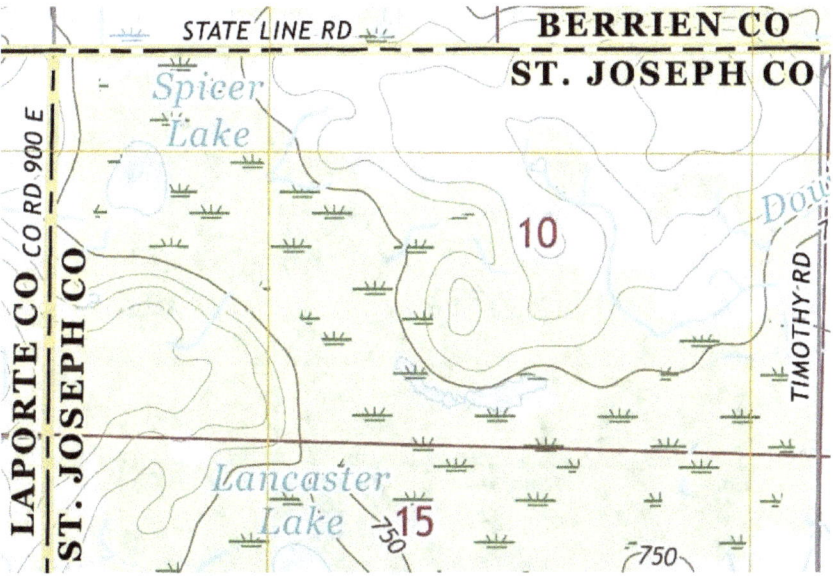

Figure 78. Spicer Lake (with a boardwalk) and Lancaster Lake are kettle lakes. (U.S. Geological Survey Three Oaks MI/IN Quadrangle, (2017)**)**

Figure 79. Spicer Lake. Note progression of three habitats away from the emergent zone at the lake. View from boardwalk.

Figure 80. Marsh from Spicer Lake boardwalk.

Little Calumet Headwaters Nature Preserve has trails, springs, upland forest, and a lake.

Springfield Fen Nature Preserve and Galena Marsh Wetland Conservation Area feature a prairie fen (spring-fed vegetation area). There are no established trails.

Cressmoor Prairie Nature Preserve contains one of the few unplowed black soil prairies that were once the most common prairies in Indiana. These rich, fertile soils are among the finest agricultural soils on Earth. Glaciers brought fresh sediment into Indiana, which weathered into great agricultural land.

1c. Kankakee Drainageways

This extensive flat land is home for the Kankakee and Tippecanoe Rivers. Glacial meltwater from nearby ice fronts spread sand over a broad area. Meltwater from the three glacial lobes (Lake Michigan, Saginaw, and Erie) converged in this drainageway.

At the height of meltwater flow, huge amounts of water poured through the region and carried sand through Indiana into Illinois. This massive flood became known as the Kankakee Flood or Torrent. The "Flood" was actually a series of large-scale meltwater drainage events about 19,000 years ago. After the waters subsided, a vast sheet of sand covered the region. Wind reworked and reshaped the sand to form significant dune fields in areas on the Indiana side as well as in Illinois (Figures 81 and 82). Flooding carried sand down the Illinois River valley and is responsible for a large sand dune field south of Peoria, IL. Thousands of sand dunes in Indiana are the result of westerly winds blowing on the vast sheet of meltwater sands.

Note: Meteorologists describe wind direction based on the direction from which the wind blows, e.g., westerly winds blow approximately from west to east.

Lakes and swamps occupied shallow low spots on the drainageways. Europeans migrated to the area and developed drainage systems to permit farming of the organic-rich, marshy soils.

The Kankakee Drainageways is bounded on the north by Valparaiso Morainal Complex, on the south by the Iroquois Till Plain and on the east by St. Joseph Drainageways, Plymouth

Morainal Complex, and Warsaw Moraines and Drainageways (Figure 2).

Figure 81. Tree-covered sand dunes on sand prairie of Kankakee Drainageways. (West of US 41 near Enos)

Figure 82. Panorama view of a sand dune on agricultural land outside preserves. An oasis of woodland habitat. Sand is mined in parts of Kankakee Drainageway.

Geologic History of the Grand Kankakee Marsh

After the Last Glacial Maximum phase (29,000-19,000 years ago) began to wane (Hughs, 2022), the Lake Michigan, Saginaw, and Huron-Erie glacial Lobes were starting to retreat. Meltwater poured into lakes that developed in back of end moraines. Drainage was initially to the south across the Tippecanoe lowlands to the Wabash River (Curry, et al., 2014). Eventually, water broke through moraines in a catastrophic rush of water called the Kankakee Torrent which flowed to the west into Illinois. The flooding events were concentrated just prior to 19,000 years ago (Kehew, et al., 2017). The flooding waters traveled so rapidly that they tore up Silurian dolostone bedrock and piled it into rubble bars, especially in Kankakee County, IL (Figures 83-85). Huge amounts of sand were spread over a vast region, forming the Grand Kankakee Marsh in a "geological instant." Sand that was swept over the area by the raging waters became the source material for later winds to scour and build innumerable sand dunes. The vast wetlands became a wildlife paradise. After Europeans discovered the "Everglades of the North," hunters from all over the world came to the area. Settlers arrived and began to drain the wetlands in the late 1800s and early 1900s. Efficient drainage systems drastically reduced the wetlands to a mere shadow of their original condition. Only about one percent of the original marsh has not been drained. Dredging of the Kankakee River provided an avenue to exit the water and wash a considerable amount of sand downstream, clogging some river channels. Bedrock just below the glacial meltwater sediments are Silurian and Devonian. A quarry in Silurian dolostone *klint* (hill of erosion-resistant fossil reef) is south of Francesville (personal communication, Steven Smith, June 2024).

Figure 83. Panorama of giant rubble bar deposited by rapidly flowing Kankakee Torrent. Bar parallels Kankakee River and stretches both directions off the photo. (Kankakee County, IL)

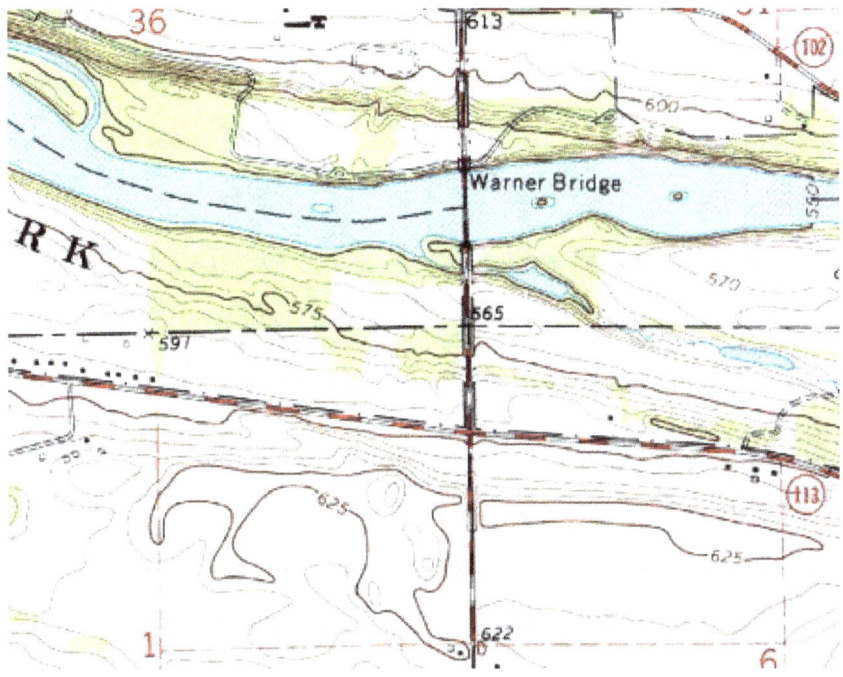

Figure 84. Topographic map of giant rubble bar (625 ft. contours) with thin layer of windblown sand on top in Figure 83. (U.S. Geological Survey Bonfield Quadrangle, IL, (1973))

Figure 85. Small rubble bar hill strewn with large gravel deposited by Kankakee Torrent. (Kankakee County, IL)

Tippecanoe River State Park

Located on the Tippecanoe River, this 2,761-acre park was established in 1943. It preserves some of the wetlands that used to dominate the area. During the height of glacial meltwater drainage, water flowed south to the Wabash River. The Tippecanoe River follows this general path through the Kankakee Drainageways. The river displays significant looping meanders (Figures 86 and 87).

Figure 86. Tippecanoe River meanders over its floodplain. Note oxbow lakes. (U.S. Geological Survey Bass Lake Quadrangle, (2016))

Geology of Indiana State Parks

Figure 87. Part of a meander loop on the Tippecanoe River. (Tippecanoe River State Park)

As you enter the park, the road passes through numerous sand dunes that are stabilized by growth of trees (Figure 88). These dunes are measured in tens of feet in height, compared to the dunes along the Indiana beaches of Lake Michigan that tower over 100 feet high. Trails allow visitors to traverse sand dunes, prairies, river bluffs, marshes, and floodplain (Figure 89). Wooded areas support a wide variety of trees.

Sand Hill Nature Preserve: This unit lies within Tippecanoe State Park. It features sand dunes formed as the wind blew over Wisconsinan glacial meltwater deposits of sand.

Tippecanoe River Nature Preserve: The floodplain displays oxbow lakes and wetlands. An abundance of wildlife and birds make their homes in the Park. See the Tippecanoe State Park brochure for details.

Figure 88. Sand dune formed by wind blowing over meltwater sand deposits. (Tippecanoe River State Park)

Figure 89. Marsh formed from a cutoff meander (oxbow lake). (Tippecanoe River State Park)

Potato Creek State Park

Potato Creek is the third most visited state park in Indiana. It was established in 1969. Lake Worster is an artificial lake of 327 acres. The park is named for the potato-like plants that were gathered along stream banks in earlier times. The area is known for many migratory birds, as well as other wildlife. Osprey nesting sites make this an unusual location.

In the 3,840-acre park, the state of Indiana has restored part of the Grand Kankakee Marsh wetlands, the largest freshwater marsh on Earth, estimated at a million acres. Agricultural use involved draining the wetlands. The state has worked to restore the land to its original pre-settlement condition. Prairie and wetlands have all received fresh infusions of native plants as a result. Numerous sites allow observation of these restored wetlands (Figure 90).

Figure 90. Restored wetland in Potato Creek State Park.

Potato Creek State Park sits on the boundary of the Kankakee Drainageways on the west and the Plymouth Morainal Complex on the east. The west's gently rolling landscape is developed on Lake Michigan Lobe clay and silt till (Wedron Formation) (Figure 91).

Figure 91. Gently rolling topography developed on Lake Michigan Lobe till of silt and clay, west side of Potato Creek State Park.

The park's east side is a more rugged landscape developed on a chaotic mixture of till, kettles, swamps, and kames, e.g., Vargo Hill. This is part of the Plymouth Morainal Complex (Figure 92).

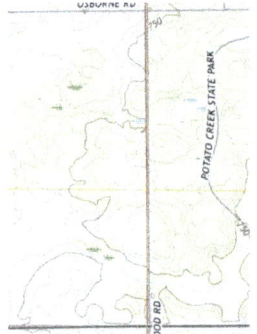 

Figure 92. Potato Creek State Park's topographic contrast is due to glacial histories of Kankakee Drainageways (west) and Plymouth Morainal Complex (east). (U.S. Geological Survey North Liberty and Lakeville Quadrangles (2016)). Interpreted from data on *Quaternary Geologic Map of Indiana***, Gray, H. H., 1989, Indiana Geological and Water Survey.**

The three glacial lobes which entered Indiana during the Wisconsinan glaciation (Lake Michigan, Saginaw, and Huron-Erie) fought over this area. If there had been just two lobes calmly touching and flowing the same direction, they might have produced ridges between the lobes. Not so here. There was so much interaction between the lobes that deposits from each lobe stacked on top of each other as they competed for space. Extensive ridges were not in the picture. Instead, till and glacial outwash sediments were deposited in a confused jumble of kames (depressions on the glacier filled with gravel/sand), kettles (pockmarks where ice blocks melted), and swamps (due to uneven deposition of glacial sediments). The chaotic collection of short ridges and poorly drained landscapes is called *hummocky topography*. Trails lead through woods and a variety of habitats. An example of restoration includes Swamp Rose Nature Preserve (Figure 93). Kames are common in the area (Figure 94).

Figure 93. Panorama of a stream that cuts through a kame at north end of Swamp Rose Nature Preserve, east side of Potato Creek State Park. (U.S. Geological Survey Lakeville Quadrangle (2016)**)**

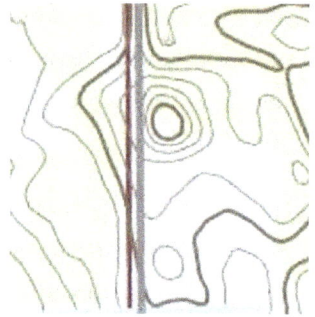 

Figure 94. Round kame hill 0.4 mile south of Potato Creek State Park. (U.S. Geological Survey Lakeville Quadrangle (2016))

Example nature preserves in Kankakee Drainageways

Kankakee Sands in Northwest Indiana and Kankakee Sands Preserve in Kankakee County, Illinois are part of the Nature Conservancy's efforts to conserve and restore part of what was once one of the largest wetlands in North America. When Europeans arrived in the region, they found wetlands estimated up to a million acres. Drainage of these wet areas for agriculture drastically changed the landscape. Nature Conservancy initiatives have focused on plant and wildlife restoration. The northwest Indiana combination of Efroymson Restoration at Kankakee Sands, Willow Slough Fish and Wildlife Area, Beaver Lake Nature Preserve, Conrad Savanna Nature Preserve and The Nature Conservancy's Conrad Station Savanna total about 20,000 acres. The nearby Nature Conservancy's Illinois segment totals about 555 acres. The protected areas include savanna, wetlands, and sand dunes, plus many plant and animal species. Kankakee Sands provides views of wet prairie and sand dunes, as well as 90 bison grazing on restored tallgrass prairie (Figures 95-97).

Kankakee Sands consists of dry, *mesic* and wet sand prairies, sand blows, sedge meadows, wetlands, and black oak savannas. There are over 86 rare threatened and endangered species. Some 600 native plant species are planted in the restorative process. More than 240 bird species are known. These include rare species such as Henslow's sparrow, northern harrier, and least bittern. The sands support 70 butterfly species, including the state-endangered regal fritillary butterfly. Over 900 species of moths are known. Dragonflies, bees, frogs, lizards, snakes, badger, and bison (Figure 102) all are present at Kankakee Sands.

Figure 95. Panorama of arc-shaped dune developed on the sand prairie and a touch of oak savanna on Kankakee Sands Preserve. (Kankakee Sands bison viewing area)

Figure 96. Bison graze on Kankakee Sands.

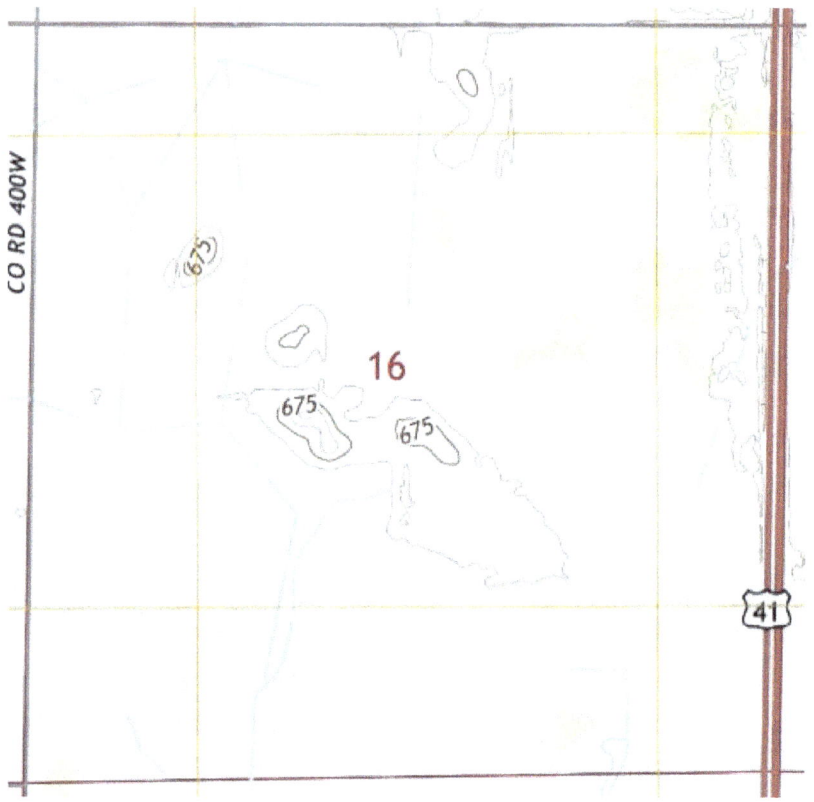

Figure 97. Sand dunes rise from sand prairie on Kankakee Sands. Figure 95 photo is of the largest 675 contour arc-shaped (parabolic) dune which indicates wind from the southwest. (U.S. Geological Survey Enos Quadrangle, (2016))

Barnes Nature Preserve: This area features sand savanna, sand forest, sand prairie, and sedge meadow. No trails. Hunting is allowed in season. Be cautious.

Conrad Savanna Nature Preserve/Conrad Station Nature Preserve: Features here include sand flats and sand dunes. Arc-shaped dunes indicate westerly winds (east side of dunes is steepest; see: https://www.in.gov/dnr/nature-preserves/files/np-ConradSavanna.pdf). Hunting is allowed in season. Be cautious.

Fisher Oak Savanna Nature Preserve: Here are located oak sand dune ridges, oak sand prairie savanna, pin oak flatwoods and high-quality black soil wet prairie.

Spinn Prairie Nature Preserve: This is one of the few remaining tracts of tallgrass prairie that once covered much of the region. No parking lot. No trails.

Tefft Savanna Nature Preserve: Features include arc-shaped sand dunes that illustrate prevailing westerly winds (steep side on the east side of dunes. No trails. Hunting is allowed in season. Be careful. Must check in. See:
https://www.in.gov/dnr/nature-preserves/files/np-Tefft_Savanna.pdf

Coastal Plain Ponds Nature Preserve: The focus here is on wetlands with plants that are found along the Atlantic and Gulf Coastal Plains. No trails. Hunting is allowed in season. Be careful. Check in is required.

Stoutsburg Savanna Nature Preserve: displays sand savanna dunes and tall grass prairie. Part is open for hunting in season. Be cautious.

1d. St. Joseph Drainageways

This small area which borders the state of Michigan is actually an extensive, low *alluvial plain* (Figure 98). Meltwater from all three glacial lobes flowed through what are now the paths of rivers in the area to the Kankakee Drainageways. Melting of ice eventually shifted drainage into the St. Joseph River. This resulted in the bend at South Bend. Blocks of ice were detached from the ice front and eventually melted, leaving depressions (kettles or kettle lakes). Devonian and Mississippian strata underlie the area.

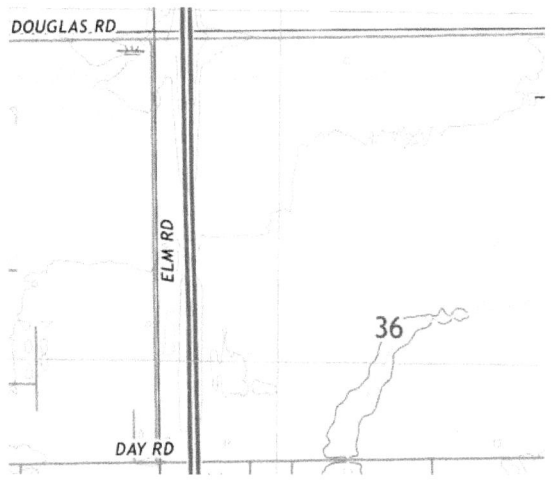

Figure 98. Flat landscape of St. Joesph Drainageways. (U.S. Geological Survey South Bend East Quadrangle, (1992)**)**

St. Joseph Drainageways is bounded on the south by the Plymouth Morainal Complex, on the west by the Kankakee Drainageways, and on the east by the Warsaw Moraines and Drainageways (Figure 2).

There are no state parks in St. Joseph Drainageways.

Example nature preserves in St. Joseph Drainageways

Boot Lake Nature Preserve has wetland, forest, and prairie habitats that surround kettle lakes (Figure 99). Sediment and vegetation partly fill the lakes. These lakes should eventually fill with peat and sediment. Over 200 bird species have been identified. There are trails and a boardwalk. Part of the restored prairie was once the City of Elkhart's sludge farm.

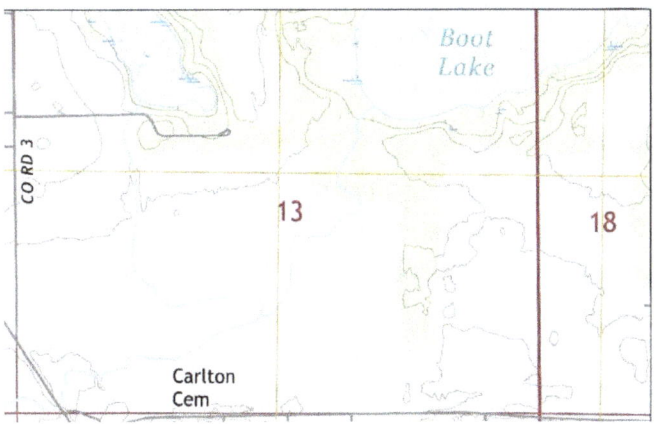

Figure 99. Boot Lake and nearby lake are kettle lakes in depressions left after isolated ice masses melted. The flat landscape is typical of St. Joseph Drainageways. (U.S. Geological Survey Osceola Quadrangle, (2016))

Fawn River Nature Preserve features the meandering Fawn River, one of Indiana's most natural rivers.

1e. Plymouth Morainal Complex

Plymouth Morainal Complex is a region where all three glacial lobes (Lake Michigan, Saginaw, and Huron-Erie) interacted. At times one or the other lobes would enter an area and deposit till and/or glacial outwash sediments. This resulted in stacking of deposits from the various lobes. The net result is a chaotic collection of short ridges and poorly drained landscapes called hummocky topography, which is undulating rough and smooth landscape that is less pronounced than knob and kettle topography. However, this landscape has markedly greater relief than the Kankakee Drainageways to the west. Numerous kettle lakes represent where blocks of ice were left behind by retreating glaciers. With time, the lakes fill with sediment and organic matter. Compacted and partly decomposed vegetation forms peat. The uneven arrangement of moraine tills and outwash formed kames (piles of sediment that accumulated on top of glaciers and were let down onto the land surface as ice melted away) as well as kettles and large kettle lakes (Figure 100).

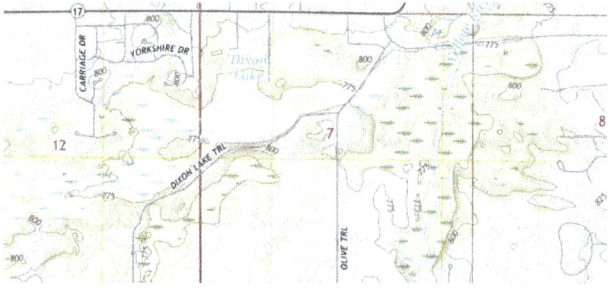

Figure 100. Uneven drainage on tills and outwash produced swamps, kettle lakes, and kames on the Plymouth Morainal Complex. (U.S. Geological Survey Plymouth Quadrangle, (2016))

Devonian shale underlies the glacial deposits and outwash sediments in this area.

The Plymouth Morainal Complex is bounded on the north by the St. Joseph Drainageways, on the east by the Warsaw Moraines and Drainageways, and on the west by the Kankakee Drainageways (Figure 2).

The only state park in this region is Potato Creek State Park. The park's eastern side is on Plymouth Morainal Complex. The western side lies on the Kankakee Drainageways. See detailed descriptions about Potato Creek in the Kankakee Drainageways section.

Lake Maxinkuckee claims to be the second largest natural lake in Indiana. Early Native Americans built mounds on the shores of the lake. Pare Mound is the largest with a diameter of 1,000 feet. It has been cut by a road. It may have been a pilot mound, used for location reference. After the mound builders left, Miami and Potawatomi groups occupied the area.

1f. Warsaw Moraines and Drainageways

The northeast-southwest trending Warsaw Moraines and Drainageways parallel the Plymouth Morainal Complex and the Auburn Morainal Complex. These physiographic areas all reflect the glacial movements and meltwater drainage influences in northern Indiana. Unlike the disorganized Plymouth Complex, the Warsaw Moraines and Drainageways has extensive, well-arranged tills and glacial meltwater sediment units. The Huron-Erie Lobe formed ridges of till with moderate relief along the southeastern border of the Warsaw Moraines and Drainageways. Devonian sedimentary rocks underlie most of this area.

To the northwest are broad, northwest-sloping glacial meltwater fans in front of the moraines. The fans may contain kettles where ice blocks separated from the main glacial mass. These blocks melted, leaving depressions. Otherwise, the fan surfaces are smooth to hummocky without well-developed drainage. Fans are thought to have been fed by glacial meltwater moving through tunnels at the base of glaciers. The Tippecanoe River now follows one of these drainage systems.

To the northeast, the Elkhart and Pigeon Rivers drain the area. This was where the Saginaw Lobe exerted control, forming a rugged, complex landscape. These rivers carried glacial meltwater to the St. Joseph River. Kettle lakes are common, and several groups form chain-of-lakes which drain from one to the next. Over time, kettle lakes tend to fill with peat and mud.

Warsaw Moraines and Drainageways is bounded on the northwest by the St. Joseph Drainageways, on the west by the Plymouth Morainal Complex and the Kankakee Drainageways, on the east by the Auburn Morainal Complex, and on the south by Bluffton Till Plain and Tipton Till Plain (Figure 2).

Warsaw Moraines and Drainageways borders Pokagon State Park. See that discussion under the Auburn Morainal Complex. There are no other state parks in Warsaw Moraines and Drainageways.

Example nature preserves in Warsaw Moraines and Drainageways

Ropchan Memorial Nature Preserve: This preserve contains moraine ridges, kettles, glacial erratics, swamps, bogs, and peat (Figure 101).

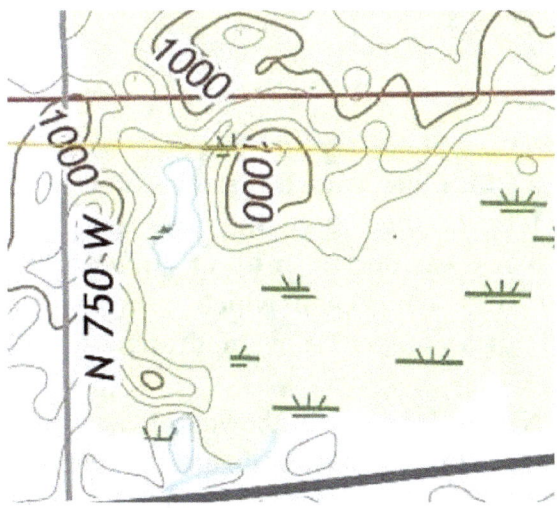

Figure 101. Ropchan Memorial Nature Preserve. Kettle lakes and swamps are apparent. (U.S. Geological Survey Orland Quadrangle, (2016))

Olin Lake Nature Preserve features Olin Lake, a kettle lake which is the largest lake in Indiana with an undeveloped shoreline (100 acres) (Figure 102). The entire shoreline is protected by this preserve. The low, marshy shoreline contains swamp forest. Upland areas have various tree species. The lake bottom is covered with marl, a white calcite precipitate from the water. Marl requires spring-fed calcareous water.

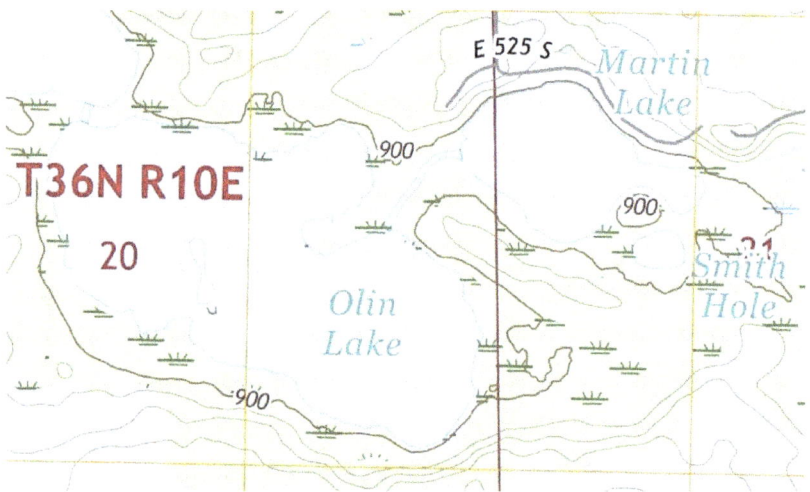

Figure 102. Olin Lake Nature Preserve. This kettle lake is the result of a large ice block breaking away from the glacial front. After the block was buried in till or outwash, the ice melted, forming a large depression which filled with water. (U.S. Geological Survey Oliver Lake Quadrangle, (2016))

Spurgeon Nature Preserve is known as "The Knobs", which is a rolling upland with small kames (Figure 103) produced by streams running on top of glaciers depositing sand and gravel in crevasses or pits on the glacial surface or near its edge. When the glacial ice melted, the deposits were lowered onto the land surface.

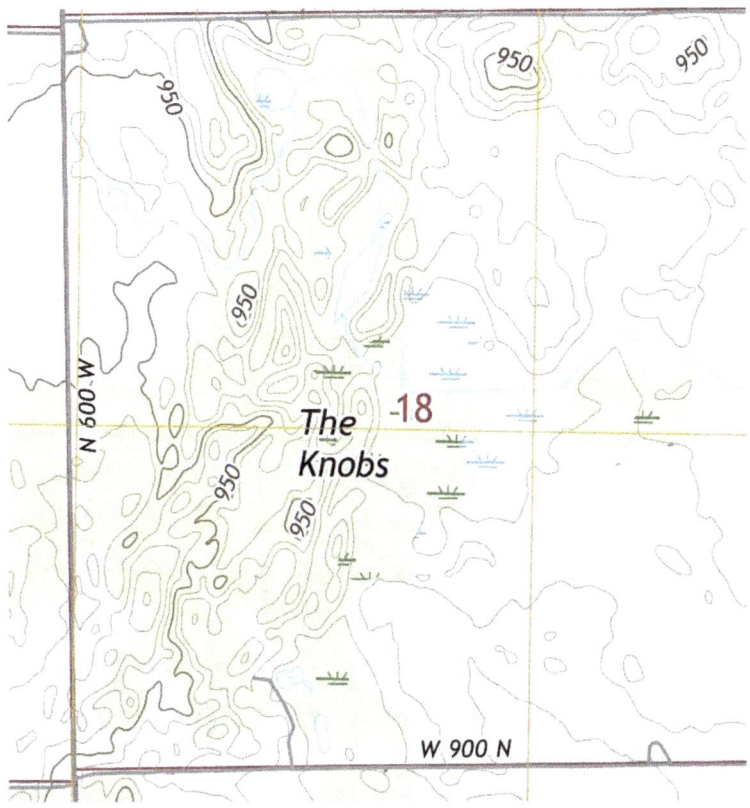

Figure 103. Spurgeon Nature Preserve. "The Knobs" are low hills (kames) which form by glacial meltwater ice-surface streams depositing sand and gravel in low spots on top of the ice sheet. Once the ice melted away, the deposit was lowered to the ground. Swampy conditions are characteristic of the chaotic processes of deposition by the melting glacial ice. (U.S. Geological Survey Ligonier quadrangle, (2016))

1g. Auburn Morainal Complex

The shape of Auburn Morainal Complex, in the northeast corner of Indiana mirrors the outline of interacting glacial lobes. The ridges of moraines have moderate relief (Figure 104).

Figure 104. Panorama of rolling landscape on a Lake Erie Lobe moraine north of Auburn. (Auburn Morainal Complex)

The Lake Erie Lobe produced the Fort Wayne and Wabash Moraines. The St. Joseph River valley lies between these ridges. With moraines-oriented northeast-southwest, glacial meltwater drained to the southwest. Later glacial changes caused the St. Joseph to join with the St. Marys River to form the Maumee River which now flows northeast and enters Lake Erie. The Eel River still flows southwest as part of the Wabash River drainage.

In the northeast part of the Auburn Complex, the Saginaw Lobe and Erie Lobes interacted with each other. When glacial lobes fuss, the result is often a hodge-podge of rugged, chaotic landscape without integrated drainage. At Pokagon State Park this is expressed where kames, eskers, and kettle lakes form a haphazard collection of features. Large kettle lakes like Sylvan Lake and Lake James stand out. Many smaller kettle lakes are scattered over the landscape. Lakes were also formed where subglacial streams cut into the land. The lack of interconnected streams is obvious in this physiographic area. Groundwater feeds some lakes. With time, sediments and vegetation began to fill the lakes, forming peat.

The Auburn Morainal Complex is bounded on the west by the Warsaw Moraines and Drainageways, on the south by the Bluffton Till Plain, and on the southeast by the Maumee Lake Plain (Figure 2).

Pokagon State Park

Pokagon State Park is an excellent example illustrating the above description of the Warsaw Moraines and Drainageways. The Saginaw and Huron-Erie Lobes interacted with each other to produce a chaotic mixture of landforms. Each lobe ran over the other and deposited till. This resulted in irregular mounds instead of smooth, U-shaped moraines found where only one lobe was involved.

Pokagon is one of the most rugged of northern Indiana's state parks. Steuben County alone has between 49 and 120 lakes, depending on the source quoted! Regardless, kettle lakes are abundant. Publicity suggests there are more lakes here than in any other county in Indiana, courtesy of its glacial history. Moraine ridges of till mark standstills of the melting glaciers. An extensive toboggan run is present on one of the steep moraines (Figure 105).

Figure 105. Panorama of toboggan slide on steep moraine. (Pokagon State Park)

The rugged landscape made farming difficult, so this is an ideal location for a park. The landscape has little changed from pre-settlement days. Erratics dropped by the glaciers as they retreated are scattered about the park (Figure 106). The rock types are not part of local Mississippian bedrock. Glaciers transported these large masses from Canada. They weathered out of glacial deposits.

Figure 106. Erratic boulders in Pokagon State Park were carried from Canada by Wisconsinan glacial lobes.

Figure 107. Lower basin of Lake James, a large kettle lake formed after a huge, disconnected block of ice separated from the glacial ice front, was buried by till or outwash, and melted to form the basin which filled with water. (Pokagon State Park)

Kettle lakes mark where blocks of ice broke off from the main ice front and later melted to form a depression where water could accumulate. Lake James is in the Seven Sisters chain of kettle lakes that surround or are near the park (Figure 107).

The Pokagon State Park sits near the last place where glacial ice resided before finally taking its leave of Indiana. Deposits are thought to date from 17,000 years ago (Indiana Geological and Water Survey (IGWS), 2018).

Potawatomi Nature Preserve sports a variety of habitats, including moraine ridges, marshes, sedge meadows, wooded swamps, and hardwood uplands. On Trail 3 that passes through Potawatomi Nature Preserve is a 135-foot overlook of this preserve known as Hell's Point (Figure 108). The name has nothing to do with the view! This high point is a kame, a mound formed by glacial meltwater depositing sand and gravel in a glacial crevasse or low spot on the glacier.

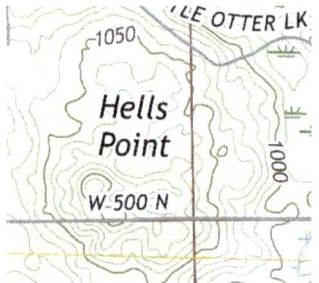 

Figure 108. Hells Point kame formed by stream deposition on top of a glacial lobe, followed by melting of the glacier, allowing gravel and sand to accumulate on the land. (U.S. Geological Survey Angola West quadrangle (2019)**) (Pokagon State Park)**

Trail 6 follows an esker (Figures 109-111). Eskers are gravel and sand deposits that filled a tunnel under the glacier. They are molds of the interior of the ice tunnel. To reach the esker, a boardwalk crosses a swamp next to Lake Lonidaw, a kettle lake (Figure 112).

Geology of Indiana State Parks

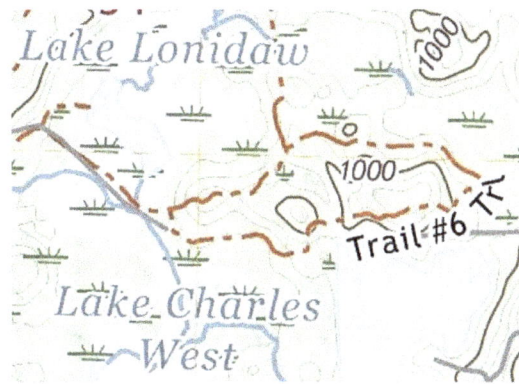

Figure 109. Trail 6 parallels an esker, a subglacial stream deposit of gravel and sand. (Pokagon State Park)

The esker's east-west orientation suggests westerly flow of glacial ice and meltwater from the Lake Erie Lobe. Lakes Lonidaw and Charles West formed in depressions left by melted ice blocks (kettle lakes). With swamps, the many features in Pokagon State Park combine to yield a chaotic arrangement with *deranged drainage* (U.S. Geological Survey Angola West Quadrangle, (2019)).

Figure 110. East-west panorama across the crest of the esker in Figure 109 where Trail #6 crosses the east end of the meandering subglacial stream deposit. (Pokagon State Park)

Figure 111. Boulders, gravel and sand make up Trail #6 esker. Pen for scale. (Pokagon State Park)

Figure 112. Marsh next to kettle Lake Lonidaw (see Figure 109). (Pokagon State Park)

Chain O' Lakes State Park

The Huron-Erie Lobe retreated to the northeast on its way toward Ohio in the waning centuries of the Wisconsinan glaciation. Stagnating ice broke off in large chunks to form *kame and kettle topography* (also called *knob and kettle topography*). The park's variety is considered one of the best in the mid-continent. Kames are mounds of sand and gravel that formed on the glacier's surface by meltwater transporting sediment into low spots or crevasses. Once the ice fully melted, the mound of sediment was lowered onto the land surface. They are the knobs. Blocks of ice between the kames melted, forming depressions for meltwater to collect to form kettle lakes.

Kettle lakes at the park form a chain between the kames (Figures 113-119). Sometimes, ice blocks were covered by glacial till or meltwater outwash sediment. This would insulate the ice and slow the melting process. Not all kettles form lakes. That depends on available water sources and drainage. Nine of the thirteen kettle lakes are connected by channels. Four are not directly connected. The interconnected lakes spawned the Chain O' Lakes name for the park.

Kettle lakes can fill with sediment and vegetation. These are described as "fresh-water bog over deep peat deposits". Peat is highly concentrated, partially decomposed vegetable matter. Peat is sometimes mined for commercial purposes. Check out your local landscape source or hardware store. Eskers are gravel and sand ridges that filled subglacial stream tunnels. Once the ice melted, the eskers remained as molds of the tunnels.

Geology of Indiana State Parks

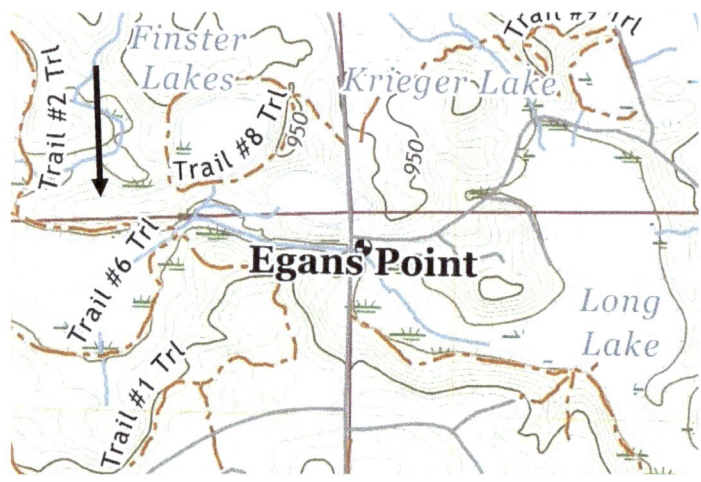

Figure 113. Kettle lakes produced by melting ice blocks. The 950 contour on Trail #8 is a kame. The elongate hill paralleling a reference line is an esker (arrow). Trail #1 parallels the south side of the esker. (Chain O' Lakes State Park) (U.S. Geological Survey Ege Quadrangle, (2022))

Figure 114. Panorama of Big Finster Lake and kame. On Figure 113: Trail #8 surrounds Finster Lake; and kame (on the right) separates kettle lakes with 950 contour. (Chain O' Lakes State Park)

Figure 115. Outwash gravel and sand typical of sediment deposited by meltwater to form kames, eskers, and other stream-related features in Big Finster Lake area. (Chain O' Lakes State Park)

Figure 116. Small kettle lake. Algae, plant material, and sediment are filling the lake basin. Smallest Finster Lake just east of 950-contour on Figure 113. (Chain O' Lakes State Park)

Figure 117. Panorama of Big Finster Lake, a kettle lake (See Figure 113). Sediment, algae, and plant material are encroaching on the lake. (Chain O' Lakes State Park)

Figure 118. Sand Lake, a large kettle lake where a block of glacial ice separated from the main glacier. The block was likely covered by glacial sediment before the ice melted. After melting, a depression was left which filled with water. (Chain O' Lakes State Park)

Other glacial deposits in the park include till, a mixture of clay, silt, sand, and gravel that results from melting of the ice front. Till is usually more common than kame and esker deposits over most glacial landscapes. The uniqueness of Chain O' Lakes State Park is the higher concentration of outwash deposits such as kames and eskers. Bluffs rise 200 feet above the river.

Figure 119. Stream connecting Dock Lake with Long Lake. Such connections between the lakes led to the name of Chain O' Lakes State Park.

Glacial Esker Nature Preserve highlights this amazing feature, considered the best example in Indiana (Figures 120 and 121). The esker is wooded with upland forest.

Figure 120. Panorama of Trail #1 esker in Glacier Esker Nature Preserve (Figure 113). (Chain O' Lakes State Park)

Figure 121. Steep slope on the side of the esker on Trail #1 in Glacier Esker Nature Preserve (Figure 113). (Chain O' Lakes State Park)

Glacial erratics are scattered over the park. Erratics are boulders that don't match the local bedrock, which is Devonian shale. Large rocks of igneous and metamorphic origin are evidence that the glacial lobe originated in Canada since that landscape has the nearest outcrops of these rocks. No local bedrock are exposed in the park. They are covered by glacial sediments. The park was established in 1960. Two hundred bird species are recognized, as well as a variety of wildlife. Glacial deposits are thought to date from about 17,000 years ago (Indiana Geological and Water Survey (IGWS), 2018).

Example nature preserves in Auburn Morainal Complex

Beechwood Nature Preserve: Visitors can walk on a boardwalk through a swamp. This site is just north of Pokagon State Park.

Trine State Recreation Area: Gentian Lake is one of the Seven Sisters chain of kettle lakes. Pokagon State Park's Lake James is in this chain, as are others adjacent to the park.

Ropchan Wildlife Refuge/Ropchan Wetland Conservation Area: This area has kames, a kettle lake, and wetlands.

Wing Haven Nature Preserve: This preserve contains glacial erratics, three of the Seven Sisters kettle lakes chain, kames, and various wetlands.

Charles McClue Reserve: This area displays rolling moraines, kettle lakes, and glacial erratics.

Marion's Woods Nature Preserve: The preserve contains a mesic upland oak-hickory forest with small wetland glacial depressions that are breeding areas for woodland frogs and salamanders.

Crooked Lake Nature Preserve: This is one of a chain of glacial lakes. It is one of the cleanest and deepest of Indiana's lakes. The preserve's shoreline is undeveloped.

Bender Nature Preserve: Displayed on the floodplain and bluffs of the Elkhart River is a floodplain swamp forest, upland ridges, and old fields reverting to natural conditions.

Merry Lea Nature Preserve: This area features a moraine lake, esker, and various wetlands, including peat bogs (Figure 122). A site that was mined for marl is also present. Twenty-four percent of Indiana was once covered by wetlands. Only three percent of the wetlands remain. There is an Environmental Learning Center and trails, but none are within the preserve itself. Permission is required to visit the preserve. Goshen College uses the area for educational purposes.

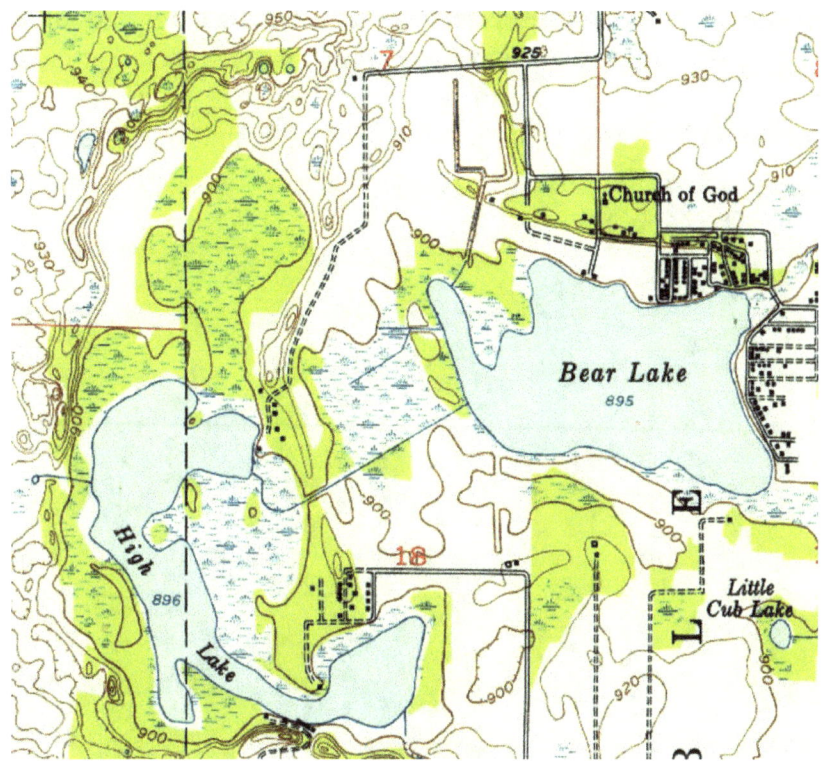

Figure 122. Merry Lea Preserve. There are several kettle lakes (depressions left by the melting of isolated blocks of ice that broke off the glacial front) and swamps. A meandering *esker* may be outlined by the 900 and 950 contours in upper left. Eskers are subglacial stream deposits of sand and gravel that are molds of an under-glacier tunnel. (U.S. Geological Survey **Ormas Quadrangle** (2016))

Lonidaw Nature Preserve: The preserve includes 3/4 of the shoreline of Little Whitford Lake, which is a five-acre kettle lake and an esker, a long, narrow ridge of coarse gravel and sand deposited by a subglacial stream. The esker indicates the under-glacier stream flowed in approximately a westerly direction (Figures 123-125).

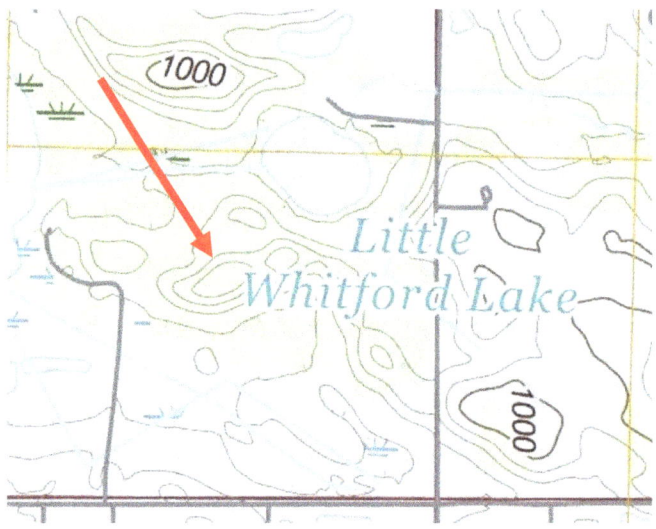

Figure 123. Little Whitford Lake is a kettle lake (hachured contours) formed after a large block of ice broke off the ice front and melted, leaving a depression. Just south of the lake is a meandering ridge of sand and gravel called an esker (arrow). This is an ice tunnel deposit by a sub-glacial meltwater stream. **(U.S. Geological Survey Corunna Quadrangle** (2016)**)**

Geology of Indiana State Parks

Figure 124. Panorama across the crest of the Lonidaw Nature Preserve esker looking southwest.

Figure 125. Panorama along the Lonidaw Nature Preserve esker ridge crest looking southeast from the lower trail.

2. Maumee Lake Plain Region

2a. Maumee Lake Plain

The last moraine ridge deposited by the Erie Lobe was the Fort Wayne Moraine. As the ice retreated, a glacial meltwater lake developed in front of the ice and behind the moraine. This was Glacial Lake Maumee, which may have eventually been the size of modern Lake Erie. A thin sheet of lake sediment was deposited over previously deposited Erie Lobe till and older lake deposits. This is the area of the Maumee Lake Plain. A few beach ridges of sand and silt let us know the elevation of the lake at various times. Bedrock below the glacial sediments is Devonian limestone.

Later, meltwater from the lake gushed through the Fort Wayne Moraine. It is thought there was a low spot where water topped the natural dam made by the moraine. This led to a massive flood downstream 17,000 years ago. This catastrophic flood has been named the Maumee Flood or Torrent. The flood event may have lasted only a few weeks (Fleming, Farlow, Argast, Grammar, & and Prezbindowski, 2018). Erosive ability of running water increases dramatically with increased speed and volume. The Maumee Torrent impacted areas downstream, scouring out a mile-wide channel that stretched for 30 miles (Indiana Geological and Water Survey (IGWS), 2018). Today, the channel is occupied by the much smaller Little River (Figure 126). There are various examples in Indiana of tiny streams occupying broad floodplains that date from the erosive activity of the Maumee Torrent. Effects of the torrent extended the entire Wabash drainage system.

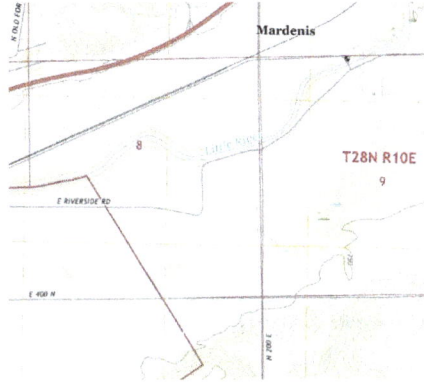

Figure 126. Little River flows on the broad floodplain valley formed by the catastrophic Maumee Torrent which rushed through a break in the Fort Wayne Moraine 17,000 years ago. (U.S. Geological Survey Huntington Quadrangle, (2016))

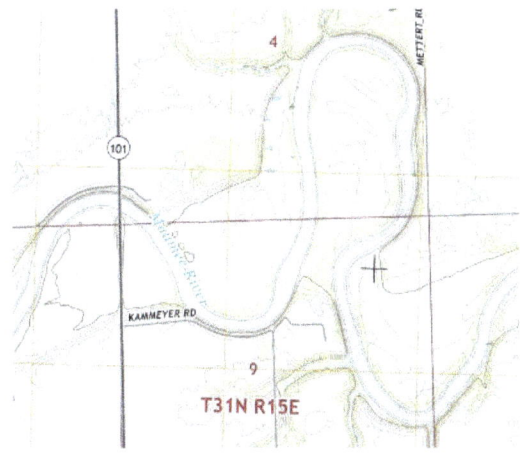

Figure 127. Wide floodplain of the east-flowing Maumee River. Floodplain rivers typically swing back and forth between steep valley walls, cutting off meanders in the process. The Maumee represents a reversal of flow in this region after Lake Maumee drained catastrophically to the west (Maumee Torrent). (U.S. Geological Survey Woodburn North Quadrangle, (2016))

The lake that developed behind the Fort Wayne Moraine is considered to be an early phase of what we now call Lake Erie. As the ice lobe receded, meltwater decreased, and lake levels dropped. As a result, the St. Joseph and St. Marys Rivers that flowed westerly reversed their courses once they joined to form the east flowing Maumee River (Figures 127).

The Maumee Lake Plain is bounded on the north by the Auburn Morainal Complex and on the south and west by the Bluffton Till Plain (Figure 2).

There are no state parks in Maumee Lake Plain.

Example nature preserve in Maumee Lake Plain

Blue Cast Spring Nature Preserve: A trail leads to views from thirty-foot-high bluffs with views of the Maumee River and an island. Spring water on this site was once sold for its medicinal value.

3. Central Till Plain Region

Till from the Huron-Erie Lobe mantles the Central Till Plain except for the southwest area which is covered by Lake Michigan Lobe tills. This vast area occupies over 40% of the state of Indiana (Gray, 2000).

3a. Bluffton Till Plain

This plain is noted for its low relief with well-defined sequences of moraines. The order for each sequence involved melting back of the glacier front, followed by thick accumulations of lake sediments, followed by readvance of the ice but not as far as the previous moraine was deposited. Each cycle resulted in clay from the previously deposited lake sediments being incorporated into new till. This produced clay-rich soils that developed on the tills (Figures 128 and 129).

Figure 128. Clay-rich soil developed on till. (Bluffton Plain)

The outermost ridge is the Union City Moraine. This is followed in sequence by the Mississinewa, Salamonie, Wabash, and Fort Wayne Moraines. The Mississinewa, Salamonie, and Wabash Rivers follow the outer edges of their namesake moraines.

The famous Late Neogene (~Pliocene) deposit in the Pipe Creek Sinkhole is a karst sinkhole formed in a Silurian reef in Grant County. This large sinkhole filled with water and sediments containing animals and plants that lived during the Pliocene. The deposit features a rare look at Indiana in the gap between the Pennsylvanian rocks and Pleistocene sediments. During Wisconsinan glaciation, tills covered the Pliocene sediments until they were uncovered during quarry operations. Silurian reefs are widely used as limestone sources due to their purity and thickness. The Pliocene sinkhole sediments yielded extinct large and small animals in this unique location (Farlow, Steinmetz, & DeChurch, 2010). Bones of extinct rhinoceros, camelids, canids, bears, rodents, and other mammals were found.

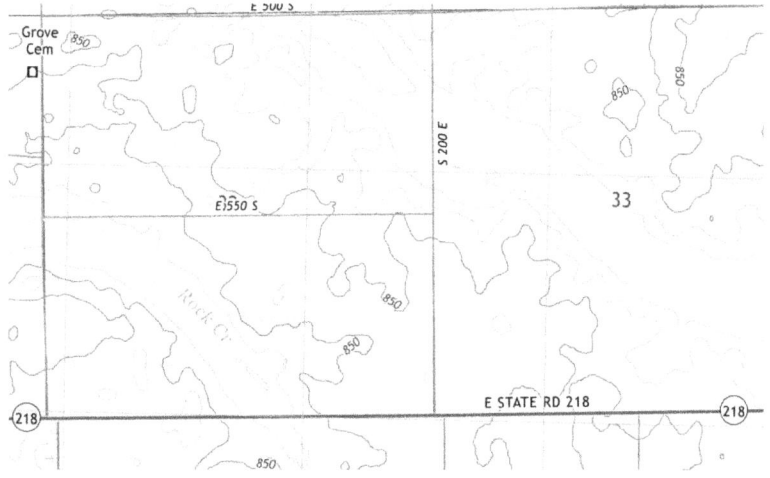

Figure 129. Low relief topography of Bluffton Till Plain. (U.S. Geological Survey Bluffton Quadrangle, (2016))

The Bluffton Till Plain is bounded by the Maumee Lake Plain on the northeast, the Auburn Morainal Complex on the north, the Warsaw Moraines and Drainageways on the northwest, the Tipton Till Plain on the west, and the New Castle Till Plains and Drainageways on the south (Figure 2).

Ouabache State Park

Ouabache State Park was established in 1962. The pronunciation is close to "Wabash" (O-bah-chee). The 1,104 acre park features a bison herd (Figure 130). The park borders the Wabash River and holds an oak-hickory forest, woodland marsh, pine plantations, and an artificial lake. Silurian limestone and dolostone underlie the glacial sediments. The area was once known for raising thousands of quail, pheasant, rabbits, and raccoons. The park name is derived from a Miami tribe's name: "water over white stones" (Figure 131).

Figure 130. Visitors can observe bison at close quarters. None were visible when this photo was taken. (Ouabache State Park)

Figure 131. Geese feeding by the lake at Ouabache State Park.

Example nature preserves in Bluffton Till Plain:

Seven Pillars Nature Preserve: Mississinewa Lake is in Miami County. The Miami tribe name means "water on a slope". The artificial lake is one of three large flood control reservoirs on the Upper Wabash River. Nearby Peru is advertised as the "Circus Capital of the World".

Erosion of cherty limestone bedrock of the Silurian Wabash Formation (Liston Creek Limestone Member) has yielded cliffs along the banks of the Mississinewa River southeast of Peru with inner chambers or caves known as the Cliffs of the Seven Pillars. Enlargement of fractures in the bedrock by the river and dissolving action of groundwater isolated the pillars and enlarged the caves (Figures 132 and 133). Located off the S. Frances Slocum Trail, the caves were used by Miami Native Americans for council meetings, social events, and as a trading post.

Seven Pillars Nature Preserve - ACRES Land Trust is on Mississinewa Road. The pillars are best viewed at the intersection of Mississinewa Road and South 400 East, west of the dam, on a short walk from the preserve parking. The pillars are not visible from the S. Frances Slocum Trail which traverses the tops of the bluff containing the Pillars.

Figure 132. Panorama of the Seven Pillars on the Mississinewa River. (Seven Pillars Nature Preserve)

Figure 133. Pillars and caves carved by stream erosion and weathering of thin-bedded limestone. Pillars are about 25 feet tall. (Seven Pillars Nature Preserve)

How did the pillars form? There are several factors necessary to produce these remarkable features. The bedrock is thin-bedded limestone. Inspect the photos and you will see there are four parallel layers of limestone. Each responds differently to the forces of erosion and weathering. Rocks vary in their resistance to erosion based on differences in composition, cementation, *permeability*, etc.

Vertical fractures called joints provide avenues for surface rainwater to penetrate the layers. Since limestone is quite soluble in water, any slight variation in the solubility of the layers is emphasized over the thousands of years these rocks have been exposed to the Earth's surface. Glacial ice covered this site about 20,000 years ago. The pillars have been forming since then.

Water seeping along joints moves laterally along *bedding planes* (parallel surfaces separating the individual beds). Solution of the rock weakens the bonds holding the layers together. This part of Indiana experiences numerous freeze-thaw cycles every year. Expansion and contraction of water in the rocks pries them apart.

Geology of Indiana State Parks

Add to these influences flooding events by the river on a regular basis. Swirling water, often carrying sediment, tends to wash away and wear down the exposed thin beds. Since major floods are less common, this effect is limited to the lower levels on the bluff.

With variations between the rock layers and the weathering and erosion effects, pillars and caves are a natural outcome. Figure 134 illustrates a possible progression in the development of pillars. Lateral variations in the rocks and spacing of joints probably account for different pillar development along the bluff face.

Figure 134. A possible progression in the development of pillars. (Seven Pillars Nature Preserve)

Hathaway Preserve at Ross Run contains impressive bedrock gorges carved by Ross Run (creek) on its way to the Wabash River. There are two waterfalls and 75-foot cliffs carved into Silurian dolostone. The rocks display ancient fossil reefs. Modern habitats include moist and dry upland forests and a floodplain forest. The waterfalls illustrate waterfall retreat on a tributary due to incision of the Wabash River by the Maumee Torrent. The tributary incised deeper to keep up with the lower river level. When a resistant layer is intersected by the tributary, a *knickpoint* (waterfall) is formed. Waterfalls retreat upstream by erosion at the knickpoint.

Salamonie River State Forest: This area features Salamonie River gorges, ravines, river bluffs, overlooks, and hills. It was chosen as a demonstration riverside forest for the reclamation of eroded land.

Kokiwanee Nature Preserve lies within Salamonie River State Forest. It features Frog Falls, Kissing Falls, and Daisy Low Falls.

Hanging Rock Nature Preserve (National Natural Landmark) is a Silurian dolostone and limestone fossil reef rising 84 feet above the Wabash River (Figure 135). It is located 1.5 miles southeast of Largo on Hanging Rock Road. The reef was exhumed after being buried under younger strata. Silurian reefs are more resistant to erosion than other limestones. Many stand above the landscape and are mined for solid limestone and dolostone. Near the end of the Pleistocene, flood waters of the Maumee Torrent raced down the Wabash River valley 17,000 years ago. A low spot in the Fort Wayne moraine allowed meltwater from the Erie Glacial Lobe to flood the area downstream. The tough dolostone and limestone reef resisted this erosion. This "pinnacle reef" is well known in Indiana and Illinois. They grew in shallow Silurian seawater at the edges of the Illinois and Michigan basins. They are called "klints" (short for "klintar", pillar-like features found on the Island of Gotland, Sweden's largest island) (Shrock, 1929). Their durability makes them stand above the surrounding landscape. Buried Silurian reefs are natural petroleum reservoirs due to their high porosity. They have been used to store natural gas in Indiana (Becker, 1974). Klement (1967) estimated that 40% of world oil comes from similar masses.

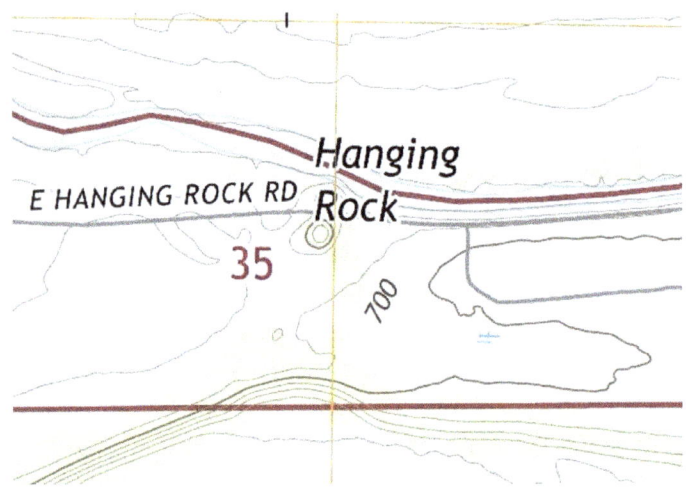

Figure 135. Hanging Rock Nature Preserve. The reef rock was once connected to strata exposed on the south bluff of the Wabash River. Due to the resistance of tough reef rock, the Maumee Torrent was not able to erode it. Instead, the torrent washed away surrounding rocks. (U.S. Geological Survey Largo Quadrangle, (2016))

The exposed Silurian Wabash Formation is divided into two members. The more resistant Liston Creek Limestone (dolostone and limestone) sits on older, more easily eroded Mississinewa Shale. Silurian reefs were constructed of calcite or aragonite secreting organisms bound together (corals, *stromatoporoids*, crinoids, cephalopods, and others). Converting calcite or aragonite to dolomite destroyed many fossils in the reef. The fossils made reefs possible and gave them durability and high porosity. No collecting of rock or fossils is allowed at the preserve.

Davis-Purdue Agricultural Center Forest: This is "the best old growth oak-hickory forest on the Till Plain and possibly one of the finest in the eastern United States". (National Park Service).

Fox Island Nature Preserve (in Fox Island County Park): This preserve is near the western outlet of Glacial Lake Maumee. Voluminous floodwaters of the Maumee Torrent (17,000 years ago) (Fleming, Farlow, Argast, Grammar, & and Prezbindowski, 2018), poured through a low part of the Fort Wayne moraine dam. Water spread a layer of sand over the area. Since then, wind has worked the sand into dunes. One dune is forty feet tall and stretches the length of the Fox Island Preserve. Marshes and swamps are present around the dune. Over 190 species of birds have been recorded in the preserve.

Eagle Marsh: This site sits on the continental *divide* between the St. Lawrence and Mississippi River drainages.

Meno-aki Nature Preserve: This area contains the easternmost occurrence of a hill prairie. Floodplain, bluffs, and ravines are present.

Tom and Jane Dustin, Robert C. and Rosella C. Johnson, and Whitehurst Nature Preserves: Cedar Creek, on the southern boundary of the Dustin Preserve is one of three **Indiana Natural, Scenic, and Recreational Rivers**. There are steep bluffs. The other rivers with this designation are Wildcat Creek in Tippecanoe and Carroll Counties and Blue River in Harrison, Crawford, and Washington Counties.

Vandolah Nature Preserve: Cedar Creek features an eighty-foot bluff view of the floodplain and stream meanders.

Loblolly Marsh Nature Preserve's creek bed pothole formed as gravel in swirling water eroded bedrock. An ADA Trail is present.

3b. Iroquois Till Plain

The Iroquois Till Plain landscape is flat and poorly drained with low Lake Michigan Lobe moraines buried by thin Huron-Erie till deposits. Where retreating glaciers leave behind a relatively uniform layer of till as opposed to significant end moraine ridges, the landscape is called ground moraine. This does not mean there is no topography. Hills, such as kames and low moraines exist. Kames are crevasse fillings or depressions on the ice which have been deposited by streams flowing on top of a glacier. After the glacier melted away, the gravel and sand deposit was laid down on the land surface (Figure 136). Weathering of till develops productive soils (Figure 137).

Figure 136. Gravel Hill, a kame, rises above the Iroquois Till Plain. This kame was laid down as the Huron-Erie Lobe glacier melted back, setting down sand and gravel that had accumulated in a crevasse or low spot on the ice. The hill's height favors wind generation. (US 41 just south of the junction with US 52)

Figure 137. Thin layer of Wisconsinan glacial till deposited by the Huron-Erie Lobe on tilted and faulted bedrock at the Kentland Impact Disturbance quarry. A soil developed on the till, permitting agricultural use of the land. (Iroquois Till Plain)

Except for the Kentland Impact site, bedrock under the glacial deposits is mainly Mississippian with some areas of Devonian and Pennsylvanian strata. The extraterrestrial impact disturbed local bedrocks.

The Iroquois Till Plain is bounded on the north by the well-defined north edge of the Iroquois Moraine, on the southeast by the northwest valley wall of the Wabash River (adjacent to the Tipton Till Plain) and the eastern side grades into Kankakee Drainageways where a sheet of sand and lake sediment rest on top of till (Figure 2).

There are no state parks and a dearth of nature preserves on the Iroquois Till Plain.

3c. Tipton Till Plain

The Tipton Till Plain is one of the largest of the physiographic regions of Indiana. It is marked by very low relief. The uniform landscape has only a few meltwater fans, kames, and moraines on the southwestern and western edges. The ice moved across a flat surface, stopped, and melted without the usual buildup of significant ridge moraines. This is a ground moraine of till where there are no significant end moraine ridges. Instead, the glacier left some ice disintegration features, e.g., kames. Kames form where gravel and sand accumulate in a depression on the glacial ice. As ice melted away, the sediment was let down onto the land surface. Good drainage allows the development of the best agricultural soils in the state on the rich soil developed from till.

Bedrock underlying glacial deposits is mainly Silurian, Mississippian, Devonian, and some Ordovician sedimentary rocks.

The Tipton Till Plain is bounded on the northwest by the northwest valley wall of the Wabash River, on the northeast by the Union City Moraine with its more clay-rich soils (at the Bluffton Till Plain), and on the southeast by the northwest valley wall of the White River (New Castle Till Plains and Drainageways). To the north is the Warsaw Moraines and Drainageways, the Central Wabash Valley is a transitional boundary (due to erosion by the Wabash River) to the west, and the south is where the Huron-Erie tills end (Figure 2).

Prophetstown State Park

This is the newest state park in Indiana. It was established in 2004 and is located at the junction of the Tippecanoe and Wabash Rivers. The 2,000-acre park features Native American history, farming as it was in the 1920s, recreation, and restoration of grassland, wetlands, fens, and savannas. Devonian bedrock is buried under hundreds of feet of glacial deposits.

About 17,000 years ago, there was a catastrophic release of meltwater from Glacial Lake Maumee when a break occurred in the Fort Wayne glacial moraine. This resulted in flooding of the Wabash River valley to depths of fifty feet at the park site. *Terraces* on the sides of the valley were formed during stream erosion by the Maumee Torrent (Figures 138 and 139).

Figure 138. Upper terrace (an ancient floodplain of the Wabash River) in the background. The car sits on the middle terrace (a later ancient floodplain of the Wabash River). (Prophetstown State Park)

Figure 139. Panoramic view of the lower floodplain terrace of the Wabash River. (Prophetstown State Park)

<u>Example nature preserve in Tipton Till Plain</u>

Granville Sand Barrens Nature Preserve: The Maumee Flood over-topped the natural dam of the Fort Wayne moraine. Glacial outwash and sand spread across the area by the running water. A sand barrens-prairie-savanna complex present here possesses windblown sand that supports a unique dry ecosystem. Sand barrens are now rare in Indiana. Twenty-two acres of the forty-acre site are the best sand barrens in Indiana. The preserve is closed in November for deer hunting.

3d. New Castle Till Plains and Drainageways

The New Castle Till Plains and Drainageways is the largest physiographic domain in Indiana. Its distinguishing feature is the number of tunnel valleys that cross it in a south to southwest radial pattern which reflects drainage from the Huron-Erie Lobe shape. Tunnel valleys are thought to form where meltwater streams run under a glacier. Tunnel valleys drain into the White River, the East Fork of the White River, and the Whitewater River. Criss-cross patterns of drainage distinguish the New Castle from the Tipton Till Plains. Otherwise, both are relatively featureless (Figure 140).

Bedrock buried under the glacial deposits includes strata of the Devonian, Silurian, and Ordovician Periods (Figure 141). The southern limit of glaciation is marked by thin glacial sediments with bedrock showing through to the surface of the plain. Erosion since the glacier receded from its greatest extent accounts for thinning of glacial sediments at the edge. The southeast part grades into the Southern Hills and Lowlands Region.

The boundaries of the New Castle Till Plains and Drainageways are clearly defined. The northwest valley wall of the White River marks the northwest boundary with the Tipton Till Plain, the north boundary with Bluffton Till Plain is marked by soils that are more clay-rich, and the southern boundary is marked by the limit of Huron-Erie Lobe till (Figure 2).

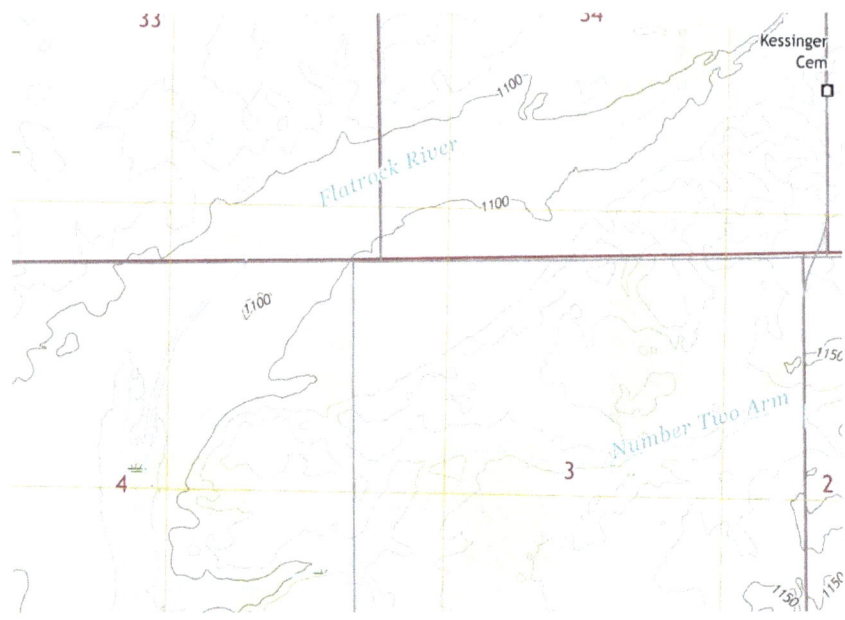

Figure 140. Featureless New Castle Till Plains and Drainageways landscape. (U.S. Geological Survey New Castle East Quadrangle, (2016))

Figure 141. Devonian Geneva Dolomite Member of the Jeffersonville Limestone overlies shale exposed in an IN-46 roadcut west of Hartsville.

Mounds State Park

This Native American site was built during various times with major construction about 160 A.D. by the Adena people and slightly later by the Hopewell (both are Woodland Period hunter-gatherers). The 252-acre park features ten unique earthworks. There are 300 prehistoric earthworks known in central Indiana. These include mounds, circular enclosures, and complexes, examples of which are found in the park (Figures 142-145).

Figure 142. Panorama of "Circular Mound," a horseshoe-shaped earthwork constructed about 1 A.D. The gateway to the mound is aligned to the east. The gateway to the center of the feature is aligned to the rising of the sun on the spring and fall equinoxes (March 21 and September 21). Photo was taken at the front of the mound looking west. (Mounds State Park)

Figure 143. Map of Circle Mound. (Mounds State Park brochure)

Figure 144. Panorama of Great Mound, aligned with gateway to the earthwork center. Ceremonies were likely held in the center of the mound with observers on the outer ring with the circular moat separating them. Post holes were first constructed and aligned with rising stars and seasons about 250 B.C. The embankment was built about 160 B.C. Dips and rises on the embankment marked events in the sky. A log tomb was constructed in a small mound on the center platform about 50 A.D. Burials were not the focus of the Great Mound.

Figure 145. Side view of the Great Mound.

Figure 146. Gravel bar on the White River. (Mounds State Park)

The site was built on the banks of the White River (Figure 146). Incision by the river exposed several small caves in the Silurian Wabash limestone. These were filled in or dynamited shut for safety purposes in the 1920s. The park's features include limestone bluffs, upland woods, floodplain woods, creek ravines, wet meadows, fens, and seeps (springs).

As is true for so many parks and nature preserves, it was the foresight and zeal of one or more individuals who guarded or lobbied for protection of the site so that future generations could learn and enjoy the features. The Bronnenberg family guarded the mounds to discourage *artifact* hunters (Figures 147 and 148).

Figure 147. Frederick Bronnenberg Jr. built this house with hand-made bricks. Silurian Wabash limestone from the banks of the White River serve as foundation. (Mounds State Park)

Figure 148. Bronnenberg house built with hand-made bricks and Silurian Wabash limestone foundation. (Mounds State Park)

Summit Lake State Park

This is a 2,680-acre park of which 800 acres are occupied by an artificial lake constructed to back up the Big Blue River. The lake was designed for flood control and recreational purposes. It was established in 1988 on the highest point in the immediate area and is near the highest elevation in Indiana. The actual highest point is Hoosier Hill north of Richmond at 1,257 feet above sea level.

The park is known for its woodlands, rolling hills, scenic valley, upland hardwood forests, wetlands, and the largest migrating waterfowl concentration in the Midwest outside of the Great Lakes (Figure 149). One hundred bird species have been recorded in the park. Pleistocene fossils of mastodon, elk, and bison are present in the area's Wisconsinan deposits. Bedrock below the Erie Lobe glacial till is Silurian. The **Zeigler Woods Nature Preserve** is located at the southwest end of the park.

Figure 149. Summit Lake.

That glaciers did their work of carrying enormous amounts of sediment of all sizes is illustrated by glacial erratics found in the area (Figure 150).

Figure 150. This metamorphic erratic boulder is a rock type not found in the local bedrock and is evidence that glaciers transported it from Canada. (near Summit Lake State Park)

The present-day small stream called the Big Blue River is nestled within a half-mile wide valley (Figures 151-154). This is an example of an *underfit stream*. The Big Blue is not the primary cause of the broad floodplain valley. An event dating from 17,000 years ago was responsible for the size of the valley. The Erie Lobe was retreating during a warming period. After depositing the Fort Wayne moraine ridge, water built up in back of the natural dam. Meltwaters from the receding glacier poured into the lake until lake levels rose high enough to overtop the moraine. These waters raced down the lowland to the southwest. The flow of water was so great that massive amounts of erosion swept sediment downstream scouring out Big Blue's valley. This catastrophic flooding is called the Maumee Flood or Torrent, named for the first glacial lake behind the Fort Wayne moraine. If you stand at the edge of Big Blue's floodplain, try to imagine a megaflood filling the valley as waters dash through and erode the valley walls. Such floods were not uncommon as glacial lobes retreated during times of warming climates. The Kankakee Torrent had already taken its toll on the landscape in northwestern Indiana and northeastern Illinois 19,000 years ago (see discussion under the Kankakee Drainageways). Even more massive flooding is recorded in Washington, Oregon, and Idaho during the slightly younger Missoula Flood.

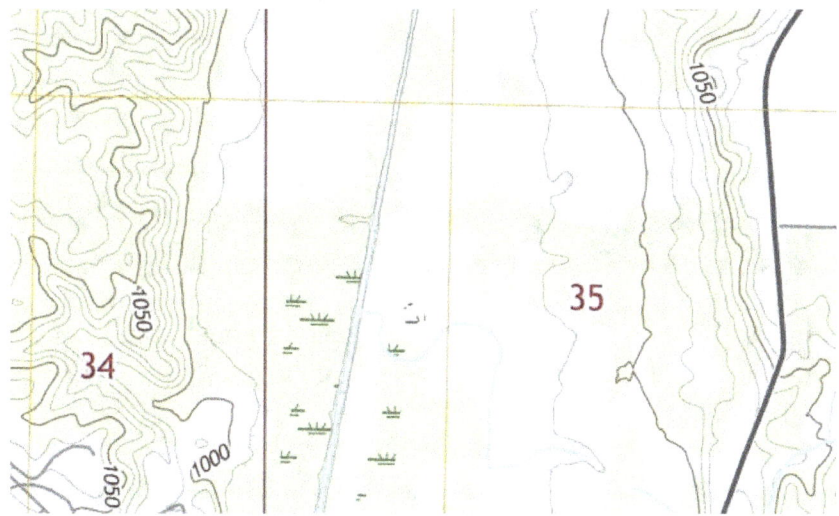

Figure 151. Big Blue River floodplain. The river is channelized to run straight. The stream is very small compared to the valley width because of catastrophic flooding by the Maumee Torrent. (U.S. Geological Survey New Castle East quadrangle, (2016))

Big Blue River is a tributary of the Upper White River East Fork. Relief is up to 100 feet adjacent to the glacial meltwater sluiceway.

Figure 152. Big Blue River valley carved by meltwaters of the Maumee Torrent. Big Blue River is marked by trees. (Outside Summit State Park)

Figure 153. East bluff of Big Blue River valley outside Summit Lake State Park. The river is in the distance to the right.

Geology of Indiana State Parks

Figure 154. Big Blue River with east bluff of the glacial sluiceway in the far distance to the left. This small stream flows in a wide valley carved by the Maumee Torrent. (West of Summit Lake State Park)

Whitewater Memorial State Park

This park was established in 1949. Its 1,710 acres is near the southernmost extent of the Wisconsinan Erie Lobe. The lake is artificial (Figure 155).

Figure 155. Locust trees blossom in the spring on the shores of Whitewater Memorial State Park.

Illinoian and Wisconsinan glaciers once covered the area. Red Springs Trail is named for a mineral-rich seep which deposits iron-bearing minerals. Ordovician strata underlie the park.

Example nature preserves in New Castle Till Plains and Drainageways

Hornbeam Nature Preserve lies along the edge of Whitewater Lake. The bedrock is like Ordovician shales and limestones seen in the exposure of **Brookville Lake Spillway**.

Brookville Lake Dam Spillway displays an extraordinary exposure of Late Ordovician Dillsboro Formation of extremely fossiliferous shales and limestones at the spillway Figures 156 and 157). These 180 vertical feet of strata contain an amazing array of marine fossils. Alternating storm deposits quickly buried marine invertebrates in the Ordovician seas, allowing excellent preservation of brachiopods, corals, trilobites, and bryozoans (see IGWS, (2018) for details about this exposure).

Figure 156. Ordovician Dillsboro shales and limestones. Thin layers record alternating storm deposits that buried invertebrate fossils quickly to preserve the shells in great detail. (Brookville Lake Dam Spillway)

Figure 157. Vultures keep watch over Brookville Lake Spillway and spectacular exposure of Ordovician storm deposits.

No fossil collecting is allowed in state parks, but loose fossils can be picked up in the city of Richmond's **Whitewater Valley Gorge Park**. Don't collect from valley walls. Fossils in the shale and limestone are evidence of the sea 440-450 million years ago during Ordovician time (Figures 158 and 159). Abundant sea life attests to warm, tropical marine environments. The Whitewater Formation is younger than the Dillsboro Formation at Brookville Lake spillway. **Thistlethwaite Falls** is the work of Timothy Thistlethwaite and his brother-in-law Joseph Ratliff in 1854. To use the Whitewater River elevation change, they dammed the water using teams of farm horses to place boulders and various stones to fill in the East Fork of the Whitewater River channel. This changed the course of the river to flow into a new channel over a rocky ledge of Ordovician thin-bedded, shaley limestone and limy shale as a waterfall. A sawmill was at the edge of the falls. A lock upstream regulated flow to turn a water wheel. The 47 feet of fall powered a grist mill, flour mill, paper mill, and lumber sawmill. None of these features remain today. The ten-foot-tall waterfall may be visited at a parking area just off Waterfall Road (Figures 160-161).

Figure 158. Ordovician Whitewater fossils are found in loose rocks on Whitewater Valley Gorge Park canyon floor. Right: brachiopod shells (*Vinlandostrophia sp.*, formerly *Playstrophia sp.*) and bryozoan colonies).

Figure 159. Ordovician fossils can be collected from loose rocks on the canyon floor, but not the canyon walls.

Figure 160. Thistlethwaite Falls, Richmond Whitewater Valley Gorge Park. Differences in erodibility of thin-bedded shaly limestone and shale form waterfall steps.

Figure 161. Panorama illustrates the erosive power of flowing water in forming Whitewater Valley Gorge canyon.

White River State Park

This 250-acre park within the city of Indianapolis was established in 1979. Advertised as America's only cultural urban state park, visitors can experience greenspaces, trails, trees, and waterways. These co-mingle with cultural, educational, and recreational attractions. Such experiences are found in the TCU Amphitheater, IMAX theater, Indianapolis Zoo, Indiana State Museum, Eiteljorg Museum, NCAA Hall of Champions, Central Canal, and Victory Field where Indianapolis Indians baseball is played. Glacial till of the Erie Lobe overlies Devonian strata.

Fort Harrison State Park

Fort Harrison State Park was established in 1996. Prior to this, the 1,700 acres was a military fort. During World War II, this was the largest reception center in the United States for persons entering military service. The park affords recreational and historical opportunities with nature preserves. Silurian limestone/dolostone and Devonian shale underlie the area covered by Wisconsinan stream deposits. This is the site of the biggest sledding hill (artificially constructed) in the area (Figure 162). The park parallels the bluffs of Fall Creek (Figure 163).

Figure 162. Sledding Hill. (Fort Harrison State Park)

Figure 163. Fall Creek as it exits Bluffs of Fall Creek Nature Preserve. (Fort Harrison State Park)

Example nature preserves in New Castle Till Plains and Drainageways

Anderson Falls Nature Preserve features a fourteen-foot-tall waterfall that is 100 feet wide along CR 1140. The resistant Devonian Geneva Dolomite Member of the Jeffersonville Limestone overlies easily eroded, fossil-rich Silurian Waldron Shale. Both these rocks accumulated in the shallow seas covering Indiana. There is a significant erosional break (unconformity) between these rocks. This unconformity records deformation in the Appalachian Mountains and those stresses caused uplift and erosion in Indiana. Never climb waterfalls. Safely view the falls from the parking lot side overlook (Figures 164-167).

Figure 164. Anderson Falls. The lip of the falls is held up by the resistant Devonian Geneva Dolomite Member of the Jeffersonville Limestone. The weaker, underlying Silurian Waldron Shale is more easily eroded. (Anderson Falls Nature Preserve)

Geology of Indiana State Parks

Figure 165. Upstream from Anderson Falls lip is a thin resistant layer within the Geneva Dolomite that forms a step in the stream profile.

Figure 166. Blocks of Geneva Dolomite break along fractures and fall after the underlying shale support is washed away.

Figure 167. Depressions in the Geneva Dolomite form by stream water dissolving slightly more soluble parts of the rock.

Ritchey Woods Nature Preserve: This preserve is part of a larger area restored to wetlands, prairies, and mesic upland and wet-mesic floodplain forests for outdoor education by the Fishers Parks Department.

Stout Woods Nature Preserve: This is a high-quality example of a Central Till Plain Flatwoods, a forest type that once covered vast areas of central Indiana with wetlands.

3e. Central Wabash Valley

Northern and eastern boundaries of this section are gradational and marked by stream dissection of the till plain draining into the Wabash River. The southern boundary is distinctly defined by the southern limit of Wisconsinan till. All tills were deposited by the Lake Michigan Lobe. The land surface is flat with low relief except where glacial meltwater erosion has done its work. The Wabash River and its tributaries have cut as deep as 200 feet into bedrock creating scenic gorges (Figures 168-169).

Figure 168. Tributary of Sugar Creek has incised Pennsylvanian Mansfield sandstone, creating a steep gorge. Sugar Creek is tributary to the Wabash River. (Turkey Run State Park) (Photo courtesy of M.S. Stillman)

Curry et al. (2018), consider the maximum extent of the Lake Michigan Lobe at 28,000 years ago. Although Erie Lobe ice did not extend into the Central Wabash Valley, meltwaters poured in the Wabash after retreat of Erie Lobe (Loope et al. (2018), date the maximum extent of Erie Lobe glaciation in central Indiana at 24,000 years ago; after melting back, the Eric Lobe readvanced 22,000 years ago). Recession began about 21,000 years ago (Indiana Geological and Water Survey (IGWS), 2018). These meltwaters rushed into the Wabash and other rivers.

During advancing phases of Indiana's three Pleistocene glacial lobes, complex interactions took place in the northcentral and northeastern physiographic areas. During warming phases, the lobes each began their retreat. End moraines were laid down by the lobes. These ranged from the outermost Valparaiso Moraine to the innermost Fort Wayne Moraine before leaving Indiana free of ice lobes.

During retreat phases, huge flooding events marked overtopping of moraines or breaks in these ridges. Water poured through the Kankakee Drainageways during the Kankakee Torrent around 19,000 years ago. Later, flood drainage through the Fort Wayne Moraine as the Maumee Torrent (about 17,000 years ago) impacted southwestern Indiana as floodwaters followed the Wabash River.

Rushing Maumee Torrent waters did significant erosion in the Central Wabash Valley. The torrent scoured the upstream channel and continued deepening the downstream section of the Wabash River. Tributary streams had to increase their downcutting into glacial tills and bedrock as the Wabash lowered its channel. These incisions produced spectacular gorges and erosional features in Turkey Run and Shades State Parks as well as various nature preserves.

The Central Wabash Valley is bounded on the north by the Iroquois Till Plain, on the east by the Tipton Till Plain, and on the south by the Wabash Lowland (Figure 2).

Figure 169. Sugar Creek, a tributary to the Wabash River, cut deeply into the Pennsylvanian Mansfield sandstone to maintain pace with the parent stream. (Turkey Run State Park) (Photo courtesy of M.S. Stillman)

Turkey Run State Park

Turkey Run is a premier park in the Indiana system in terms of scenic beauty and ruggedness of hiking trails. Due to the character of many Turkey Run's trails, hikers should be aware of the physically demanding requirements to traverse these paths. (Figures 170-172).

Figure 170. Ladders assist hikers on Trail 3 in Turkey Run State Park. (Photo courtesy of M.S. Stillman)

Geology of Indiana State Parks

As the second oldest park in Indiana, Turkey Run was established in 1916. This 2,382-acre area was spared the ravages of those who wanted the timber and any significant development by becoming a state park. As stories go for so many state parks, it was the dogged determination of private individuals who spurred the saving of these jewels of God's creative activity so everyone could enjoy them. The park is named for turkeys that occupied the area. The park's glacial history is partly revealed by a thin deposit of till coating the region's landscape. Pennsylvanian Mansfield sandstone is the bedrock exposed in the park.

Figure 171. 140 steps add interest to Trail 9 in Turkey Run State Park. (Photo courtesy of M.S. Stillman)

Rocky Hollow Falls Canyon Nature Preserve (National Natural Landmark) is the site in Turkey Run State Park with the most challenging trails. Bluffs 100 ft. high expose sandstone of the Pennsylvanian Mansfield Formation. Sugar Creek was one of the streams that worked hard to match downcutting by the Wabash River as end-of-glaciation meltwaters poured from retreating glacial lobes. The Maumee Torrent that resulted from overtopping of the Fort Wayne moraine 17,000 years ago contributed to erosion of the Wabash valley. Tributaries to Sugar Creek have also carved deep narrow channels (Figure 173).

Figure 172. The 200-foot-long suspension bridge leads to Rocky Hollow Falls Canyon Nature Preserve where most trails of Turkey Run State Park are located. (Photo courtesy of M.S. Stillman)

The Mansfield sandstone is the earliest Pennsylvanian sedimentary rock found in Indiana. It sits directly over Mississippian rocks. Their boundary is an unconformity, marking exposure of the land to stream erosion following the lowering of sea level prior to Pennsylvanian deposition. The sandstone, with some conglomerate and shale, is the product of a *braided river* system from the northeast, likely from the northern and central Appalachian Mountains and southeastern Canada (Potter & Siever, 1956). Coal present in the strata hints that ancient tropical swamp vegetation accumulated in the delta. Later in the Pennsylvanian, these delta environments yielded additional coal swamps. Three hundred million years later, Indiana industry would be partially powered by mining these coal layers. The local coal mine in the park has been closed, but outcroppings of coal can be seen in rocky walls. No collecting of anything is permitted in Indiana state parks. The Mansfield is resistive to erosion, which required Sugar Creek and tributaries to cut deep into the rock but did not form a wide floodplain which occurs in softer rocks (Figure 173). Erosion of local till resulted in glacial gravel and erratics falling into the tributaries. Rapidly flowing water caught glacial pebbles in a swirling action to carve out potholes (Figure 174).

Figure 173. Deeply incised meandering tributary to Sugar Creek swings back and forth as water erodes the outside of the curve as it also cuts down. (Photo courtesy of M.S. Stillman)

The walls of gorges reveal thin layers of shale or poorly cemented sandstone that form recesses due to their weak resistance to weathering and erosion (Figure 175).

Figure 174. Pothole bored by large gravel swirling around on bedrock sandstone. (Photo courtesy of M.S. Stillman)

Geology of Indiana State Parks

Figure 175. Alternating layers of sandstone with different resistance to erosion. (Photo courtesy of M.S. Stillman)

Waterfalls are abundant in Turkey Run. Erosion-resistant sandstone layers form the lip of waterfalls. Weakly cemented sandstone or shaly rocks are recessed below the resistant layer (Figure 176).

Weathering and erosion have ways of accentuating weaker parts of a rock mass. This results in local names for eroded features that superficially resemble a goose, punchbowl, etc.

The high iron content in Mansfield sandstone is seen in distinctive red-orange nodules and layers. Groundwater moving through the rocks deposited the iron. Some parts of southwest Indiana had such high iron concentrations that the rocks were mined during earlier times. The Mansfield was also mined as a glass sand. Iron in its chemically reduced form was used to color glass bottles green.

Figure 176. Well-cemented sandstone layer holds up this waterfall. Below are weakly resistant strata. (Photo courtesy of M.S. Stillman)

When you walk over the suspension bridge, take a moment to study the bottom of the stream. The sand comes mainly from weathering and erosion of the bedrock sandstone. Note ripples on the stream bottom. Loose sand tends to flow down slopes of ripples created by the moving water. The shape of sand ripples depends on the speed of water flowing over the sand. If you examined a cross-section of a ripple, you could see tilted layers of sand. Now consider the cross-bedding seen in the walls of the gorge (Figure 177). The absolute size of the cross-beds also depends on stream flow. Geologists use such observations to estimate directions of ancient streams flowing hundreds of millions of years ago!

Figure 177. Cross-bedding in Mansfield sandstone. Iron oxide is more resistant to weathering and is concentrated along lines in cross-beds to emphasize this effect. Current flow was from right to left. (Turkey Run State Park)

The stream bed of Sugar Creek also contains gravel washed in from till. Fossils and geodes appear to be derived from eroded Mississippian Sanders Group limestones exposed upstream (personal communication, Steven Smith, June, 2014).

Natural fractures in sandstone (joints) provide avenues for water to enter the rock and dissolve cement that holds the grains together. In winter, water freezes and pries rock masses apart. Blocks near stream channels fall from bluffs (Figure 178). Vertical faces in canyons are joint faces where blocks of rock fell (Figure 179).

Figure 178. Blocks fell from bluff by weathering and frost wedging on joints (trees grow in joint). (Photo courtesy of M.S. Stillman)

Geology of Indiana State Parks

Figure 179. Vertical joint face from which a block of rock fell. (Photo courtesy of M.S. Stillman)

Shades State Park

Upstream from Turkey Run State Park is the Shades of Death, a 3,082-acre collection of trees and gorges with a remote history of murders and myth. Don't let this fool you. The park has been called "Turkey Run without the crowds". The same Pennsylvanian Mansfield sandstone as seen in Turkey Run holds up high cliffs while the weaker Mississippian Borden Group shales and siltstones occupy the underlying exposed bedrock of the gorges. Sugar Creek's gorges were significantly deepened as Wisconsinan glacial ice melted and poured down the previously formed valleys like the Wabash River. Catastrophic flooding by the Maumee Torrent lowered the Wabash valley and triggered similar lowering by Sugar Creek. The less resistant Mississippian strata account for some variation in the local landforms in contrast to Turkey Run's tough Mansfield sandstones (Figures 180 and 181).

The unconformity between Pennsylvanian Mansfield sandstone and the Mississippian Borden siltstone and shale is visible in the gorge and Silver Cascade Falls. Unconformities are erosional breaks in the rock record. Data in Droste and Keller (1989) suggest that this part of Indiana underwent a period of erosion that began about 325 million years ago. This occurred because the Mississippian shallow inland sea withdrew from Indiana and exposed the area to stream erosion. After five million years, the sea returned and received sand washed off the land. The exposed land surface on which the sand was deposited consisted of Borden siltstone and limestone. Later burial of the sand converted it into the Mansfield sandstone. The five-million-year gap in the rock record at Shades is represented by the contact between the Mansfield sandstone and Borden rocks. The sandstone was deposited in a high energy environment that destroyed most fossils except for a few plant fragments.

Figure 180. Devil's Punchbowl. This steep-sided depression resulted from the erosive action of deeply incised tributaries of Sugar Creek. (Shades State Park)

Freezing and thawing of water in the finer grained sediments yields interesting landforms, such as the convex shape of Silver Cascade. Most waterfall undersides tend to be concave.

Established in 1947, Shades offers rugged trails that remind the hiker of Turkey Run. In earlier times, the area drew attention because of springs that issued from the contact between the Mansfield and Borden rocks. The water's medicinal value was touted, and many people desired the supposed curative effects. On Trail 1 is the Devil's Punch Bowl with springs draining the Mansfield sandstone.

Shades offers only one of three canoe trips in Indiana state parks. It is twelve miles to Turkey Run by canoe.

Figure 181. This valley in Shades State Park is widening by spring water seeping through the Mansfield sandstone. Combined with undercutting on the outside of this meander, slabs of sandstone fall to the valley floor.

Adjacent to Shades is **Pine Hills Nature Preserve.** Indian Creek's meanders lead to Sugar Creek. There are four "backbones," which are flat-topped ridges that appear to be too level to result from natural forces, but they are natural. One backbone, attributed to the devil, is 100 feet high with a ridge only several feet wide.

This and other areas are especially hazardous. See the online Pine Hills Nature Preserve trail map for precautions.

Incised meanders cut into the strata forming gorges up to 125 feet deep. Honeycomb weathering of sandstone is due to dissolving of cement holding sand grains together. Since the cementing calcite or quartz is not evenly distributed, pockmarks appear on the sandstone surface. Once a "soft spot" is found by water, it tends to retain moisture, which further accelerates weathering of the depression. This honeycomb is thought to be due to weathering of fossil algae in the sandstone.

Example nature preserves in Central Wabash Valley

Fern Cliff Nature Preserve (National Natural Landmark) is a rugged sandstone landscape with ravines and cliffs incised into Pennsylvanian sandstone by glacial meltwaters. "Dangerous cliffs" signs warn of the sharp precipices. An abandoned sandstone quarry provided sand for the greenish glass used in making the first bottles of Coca-Cola. There are no trails in the preserve.

Raccoon State Recreation Area/Cecil H. Harden Lake possess rugged Pennsylvanian sandstone cliff landscape and scenic lake overlooks.

Portland Arch Nature Preserve (National Natural Landmark) features a Pennsylvanian Mansfield sandstone natural bridge opening 7.5 feet high and 15 feet wide, with the top of the arch about 40 feet tall (Figures 182 and 183). Upstream the gorge features 100-foot-tall cliffs of sandstone. Glacial meltwaters poured through Bear Creek, creating a gorge as the stream flowed toward the Wabash River. A small tributary interacted with Bear Creek to carve out the arch. Colorful cross-bedded sandstone harks back to the days when ancient, Pennsylvanian braided rivers carried sand from the northeast to the inland sea. Cross-bedding occurs where an ancient (Pennsylvanian) stream current was carrying sand to form dune shapes on the river bed. Sand grains rolled down the back side of dunes. This angular deposition process generated cross-beds. Iron deposits in the rocks makes them resistant to weathering and highlights the cross-beds. Uneven cementation of the sandstone allows chemical weathering, frost weathering, and surface erosion to form honeycomb weathering features (Figure 184).

Figure 182. Portland Arch resulted from the erosional interaction of Bear Creek and a tributary. Once the tributary breached the narrow rock peninsula separating them, the arch was born. Continued erosion by the breaching stream enlarges the opening in Pennsylvanian Mansfield sandstone. Stacked sets of cross-bedded sandstones tell the story of deposition. The middle sandstone cross-beds curve and thin as they approach the top of the lower set. Pennsylvanian stream flow was from left to right. As sand advanced over the eroded top of the bottom layer, sand grains rolled onto this eroded surface, forming the tailing effect seen here.

Geology of Indiana State Parks

Figure 183. Stream cut through to form Portland arch.

Figure 184. Honeycomb weathering in Mansfield sandstone. Slightly less cemented areas in the sandstone are easily weathered out as small depressions. (Portland Arch Natural Preserve)

Fall Creek Gorge Nature Preserve has potholes up to twenty feet across, with depths of inches to several feet carved in Pennsylvanian Mansfield sandstone. There is a small waterfall. Glacial meltwaters eroded huge potholes with gravel derived from glacial deposits. Potholes are formed when a pebble is caught in a swirling current. The abrasive gravel cuts into easily detached sandstone. A circular depression (pothole) is the result. Fall Creek is a tributary of Big Pine Creek. The preserve was closed at this writing because of vandals and mistreatment of the site.

Weiler-Leopold Nature Preserve is near the Wabash River. The Preserve has glacial erratics along the trail. The preserve is closed in November for deer hunting.

Black Rock Nature Preserve protects high quality open oak woodlands, very rare sandstone/siltstone barrens, Pennsylvanian Mansfield sandstone bedrock exposures, sandstone cliffs, and a cave. Black Rock sandstone cliff soars one hundred and ten feet above the Wabash River revealing steep ravines and springs. Sandstone is not known for solutional caves such as those formed in limestones and dolostones. Differential weathering along weaknesses in sandstones produces alcoves, overhangs, and small caves. Dark stains are due to manganese and iron oxides deposited as groundwater coming to the surface contacts the air. The southern exposure and thin acid soils slows the growth of trees. *Selaginella rupestris* (ledge spikemoss) is an S2 (state imperiled) rare species for Indiana.

In 1811, Tecumseh's warriors waited here for General William Henry Harrison's troops before the Battle of Tippecanoe. Harrison learned of the warriors' presence and traveled inland to avoid them. In 1838, the Potowatomi Trail of Death camped near this site on their 660-mile forced relocation march from Plymouth, Indiana to Osawatomie, Kansas. During the canal era, settlers fished and provided lumber for towns east and west along the Wabash and Erie Canal on the opposite side of the river.

Williamsport Falls is in the town of Williamsport. Thick Pennsylvanian Mansfield sandstone sits over weak Mississippian Borden Group siltstone/shale. This combination forms the tallest freefall waterfall in Indiana at 90 feet. This appears to be a classic Niagara-style waterfall where turbulent falling water erodes a plunge pool and wears away the underlying softer rocks. When enough soft rock is removed, the caprock sandstone is unsupported and collapses (Figure 185). This causes waterfall retreat as the falls works its way upstream from the Wabash (Figure 186).

Fall Branch is a small tributary to the Wabash River. As glaciers melted back from central Indiana, catastrophic meltwater floods (Maumee Torrent) fed the Wabash, causing it to rapidly cut into its channel. This forced Fall Branch to incise glacial sediment and resistant Mansfield sandstone until soft Bordon rocks were exposed to erosion. This triggered the formation of a waterfall where Fall Branch entered the Wabash (Figure 187). Waterfall retreat brought the falls to its present location. Today, Fall Branch is a seasonal stream with flow over the falls ranging from none during dry seasons to high flow after significant rainfall. The waterfall is best viewed after adequate precipitation. A trail leads to the waterfall retreat canyon valley.

Geology of Indiana State Parks

Figure 185. Williamsport waterfall and plunge pool. Drought resulted in little flow when this photo was taken. Note the weaker siltstone/shale of the Mississippian Borden Group below the resistant Mansfield sandstone. (Williamsport)

Figure 186. Panorama of Williamsport waterfall and Fall Branch waterfall retreat canyon. (Williamsport)

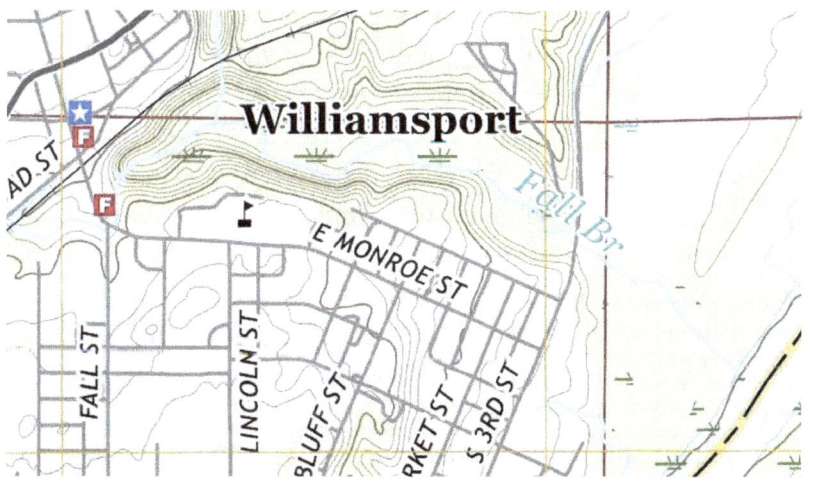

Figure 187. Waterfall retreat canyon on Falls Branch. The waterfall began at the Wabash River valley edge. (U.S. Geological Survey Williamsport Quadrangle, (2016))

4. Southern Hills and Lowlands Region

The east-west irregular boundary separating the Wisconsinan glaciation from southernmost Indiana is a composite of various ages of glacial lobe advances and is not itself a single timeline. South of this irregular line the physiographic divisions strongly reflect differences in bedrock characteristics. Some areas were never directly touched by glacial ice, while 60% of the Southern Hills and Lowlands Region experienced one or more pre-Wisconsinan glaciers. The deposits left by these more ancient glaciers are old enough to have been highly affected by stream erosion. In addition, there was a huge amount of Wisconsinan glacial meltwater carried through the area by rivers like the Wabash, White, and Ohio. These waters greatly enlarged valleys, deposited sediment, and formed river terraces. Winds formed sand dunes and loess deposits. The Maumee Torrent was a catastrophic flooding event that dramatically impacted the Wabash River drainage about 17,000 years ago. The multiple pre-Wisconsinan glacial events, stream erosion during interglacial times, plus Wisconsinan and Holocene stream action together make it difficult to unravel the detailed glacial history of southern Indiana.

The physiographic divisions that follow are arranged roughly from west to east for a reason. Bedrock outcrops in Indiana are affected by erosion of the uplifted Cincinnati Arch. The arch (Figure 25) trends southeast to northwest across the state. Strata on the southwest side of the arch tilt to the southwest (Figure 24). Erosion of these tilted strata exposes older rocks near the arch and younger rocks away from the arch. This results in bedrock exposed in southern Indiana trending from young on the west side to older on the east (Figure 188). The Southern Hills and Lowlands sections bedrock therefore range from Pennsylvanian in the west and successively older rocks through the Mississippian, Devonian, Silurian, and finally Ordovician on the eastern edge. The discussion that follows begins with the Wabash Lowland in the far southwest corner where younger Pennsylvanian rocks are seen and moves toward the east, as older rocks are successively exposed.

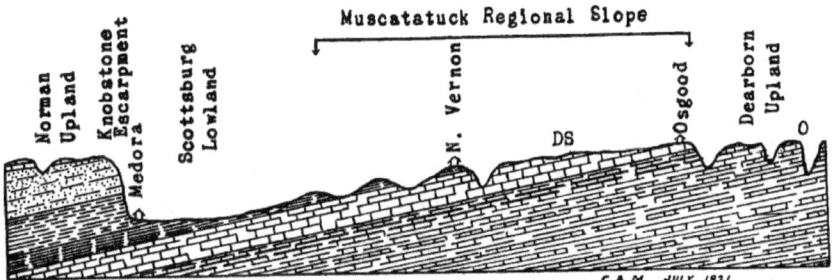

Figure 188. Topographic profile across part of the eastern side of the Southern Hills and Lowlands region. Note how the southwest dip of the rocks off the Cincinnati Arch successively exposes older rocks to the east (right) by erosion. The sketch is vertically exaggerated. From Malott, C.A., 1922, The Physiography of Indiana in Handbook of Indiana Geology. The Department of Conservation, Division of Geology, p. 160, fig. 17. Courtesy of Indiana Geological and Water Survey, Bloomington.

4a. Wabash Lowland

This southwestern part of Indiana was not touched directly by the Wisconsinan glacial ice that the northern two-thirds of the state experienced. The Lake Michigan Lobe and the Lake Erie Lobe left deposits of till to form the northern boundary of the Wabash Lowland. The only Wisconsinan deposits brought into the lowland were stream sediments, many of them carried during the Maumee Torrent event about 17,000 years ago (Figure 189). Overtopping of the Fort Wayne moraine by water from Glacial Lake Maumee allowed the torrent's effects to be experienced from near the northeast corner of Indiana to the state's southwest tip. The entire length of the Wabash River was affected by this onrush of water. Today's Wabash River is a shadow of its former self during that great flooding. Wide and deep floodplains were carved by this amazing event. The wide floodplains of the Wabash, Eel, White, and East Fork of the White Rivers possess broad terraces of Wisconsinan glacial meltwater sand and silt (alluvium).

After the torrent, sand and silt deposited on the floodplains dried out. Wind blew sand and dust off river floodplains. These sediments settled on hills near the rivers as sand dunes and loess. Sand dunes are more prevalent on the east side of the Wabash River due to the prevailing westerly winds (Figure 190).

The northern part of the Lowland is dominated by pre-Wisconsinan till plains capped by six feet of loess. Lake silts accumulated in tributary valleys. Most of the Wabash Lowland is rolling hills of low relief. The region experienced one or more of the pre-Wisconsinan glaciers. The vast amount of time separating these early glaciers and Wisconsinan drainage and wind effects allowed extensive erosion of the landscape (Figure 191). Low rolling hills lie south and east of the White River (Figure 192).

Figure 189. Lower Wabash River looking south. Sand and silt alluvium exposed on wooded floodplain edge. (Harmonie State Park)

Figure 190. Windblown loess and sand caps Wabash River bluffs rising above the floodplain north of New Harmony.

Figure 191. Rush Creek cuts through floodplain alluvium after dissecting hills adjacent to the Wabash River. (Harmonie State Park)

Figure 192. White River meanders on its broad floodplain between low hills. Note the oxbow lake. (U.S. Geological Survey Iona Quadrangle, (2016))

Pennsylvanian shale makes up the Lowland's bedrock. Shales weather easily and make a subdued landscape in contrast to regions to the east underlain by more resistant bedrock types. Coal *strip mines* are extensive in some areas. Many have been reclaimed. The presence of cyclic deposits involving coal and shale are typical of Pennsylvanian cyclothem deposits reflective of rising and falling sea levels on low coastal areas in those warm ancient times.

The Wabash Lowland is bounded on the north by the Central Wabash Valley, on the east by the Martinsville Hills and Crawford Upland, and on the southeast by the Boonville Hills (Figure 2).

Shakamak State Park

Established in 1928 to transform the devastations incurred during early 1900s coal mining, this 1,766 acre park features recreational facilities including lake fishing and boating. Shakamak is in the pre-Wisconsinan glaciated area of the northern Wabash Lowland. As was common in the days of the New Deal, Civilian Conservation Corps (CCC) workers planted trees, built roads and trails, and constructed buildings in the park.

Coal came from Pennsylvanian cyclothem deposits that reflected the ancient environment where sea level rose and fell near the shoreline in a delta landscape (Figure 193). Coal swamps of the Pennsylvanian Period accumulated vegetation in this warm, moist environment. Sea level changes were likely linked to glaciers forming and melting in the Southern Hemisphere.

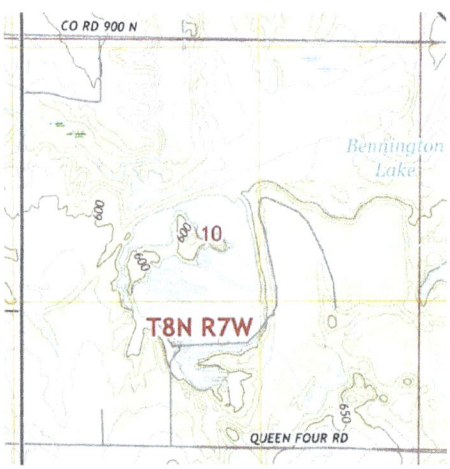

Figure 193. Example lake filling a former strip coal mine near Shakamak State Park. (U.S. Geological Survey Jasonville Quadrangle, (2016))

Prior to mining, the area was known as Golden Knobs. Three lakes now occupy the site. **Shakamak Prairie Nature Preserve** is an example of moist prairie that once covered much of the area before settlement. Shakamak in the Delaware language meant "The waters of the long fish" and referred to the Eel River.

Harmonie State Park

Harmonie State Park's 3,465 acres was established in 1966 with connections to the famous New Harmony progressive movement of the 1800s that fostered the first free public school with equal education for boys and girls, the first free public library, and the first kindergarten in the United States. New Harmony became a center for science and education and was visited by famous geologists such as Charles Lyell (Figure 194). It has been called the "Birthplace of American geology" (Indiana Geological and Water Survey (IGWS), 2018). New Harmony was the home of Thomas Say, the Father of North American Entomology, the study of insects. Say was living in New Harmony when he described Say's Firefly (*Pyractomena angulata*) in 1826. This is Indiana's state insect.

Pennsylvanian, coal-bearing, cyclothem bedrock underlies the area. Windblown loess and sand dunes derived from the Wabash River floodplain coat the landscape. Oil was discovered in Indiana in 1890. Numerous oil wells were drilled on what would become Harmonie State Park. A few still produce. An antique oil pump acts as a monument of the history of oil production in the region.

Located on the floodplain and adjacent hills of the Wabash River in far southwestern Wabash Lowland, the park is near the southernmost limit of pre-Wisconsinan glaciation in the region (Figures 195 and 196). Ravines dissect the valley walls by the river. Flooding during the Maumee Torrent and other glacial meltwaters deepened and widened the Wabash River. Adjacent tributaries had to cut deeper as a result (Figure 197).

Figure 194. New Harmony historic sign.

Figure 195. Lower Wabash River looking north. Sand and silt are exposed on valley edges. Woods dominate the floodplain. Mink Island is in the distance. (Harmonie State Park)

Figure 196. Cutbank of Bowman Bend on the Wabash River exposes silt and sand of the floodplain. (Harmonie State Park)

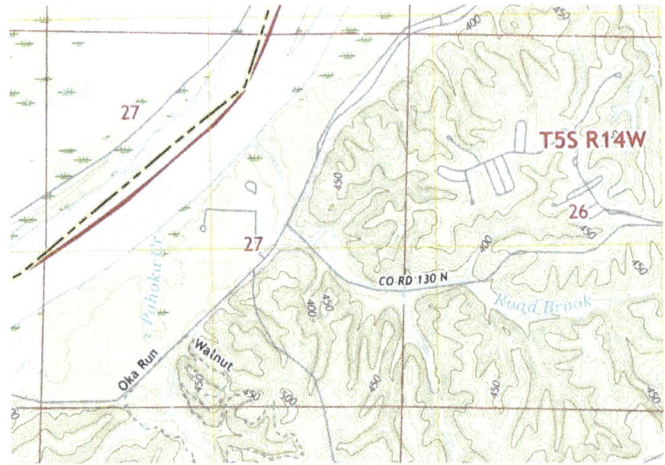

Figure 197. The topographic map of Harmonie State Park reveals extensive ravine development adjacent to the Wabash River (U.S. Geological Survey Solitude Quadrangle, (2016))

Geology of Indiana State Parks

Example nature preserves in Wabash Lowland

Near the Wabash and Ohio Rivers junction are four preserves: **Wabash Lowlands Nature Preserve, Flatwoods Nature Preserve** (a lake drained partly by sinkholes), **Twin Swamps Nature Preserve,** and **Hovey Lake Fish and Wildlife Area**. These preserves feature an array of wetland environments. Not all have trails. Some have a hunting season. Check individual websites for details. Many nature preserves are abandoned Pennsylvanian coal strip mines with lakes converted into wildlife areas (Figure 198). **Greene-Sullivan State Forest** with its rolling hills and 120 strip mine lakes display fossils in ledges around the lakes. Some areas permit hunting.

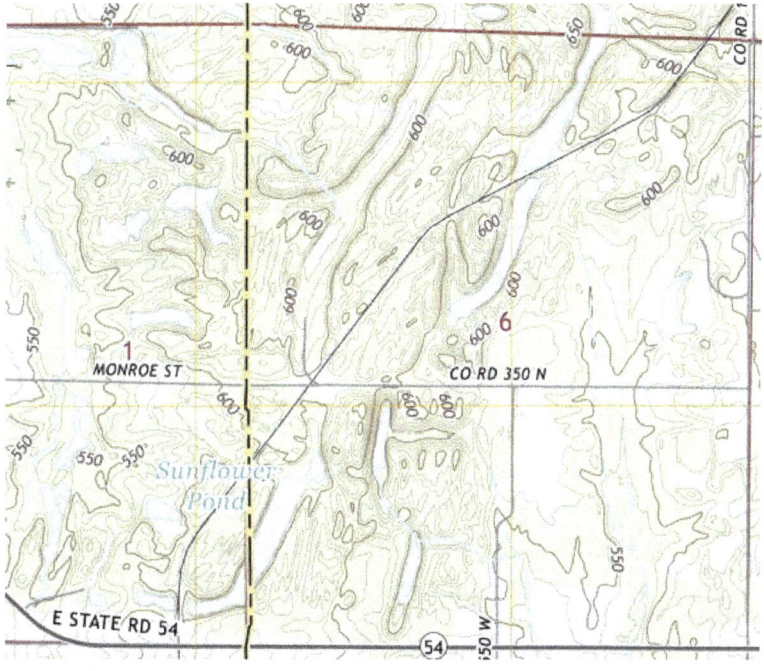

Figure 198. Strip mine lakes in Greene-Sullivan State Forest. (U.S. Geological Survey Linton Quadrangle, (2016))

4b. Boonville Hills

The Boonville Hills are distinct from the Wabash Lowland. The Hills experienced no direct glaciation, neither pre-Wisconsinan nor Wisconsinan. The lack of glacial sediment means that vegetation is closely tied to local bedrock type. It is underlain by Pennsylvanian strata of sandstone and limestone and is more hilly than the younger Pennsylvanian cyclothems dominated by weak shale and coal to the west in the Wabash Lowland. Erosion produces a more rolling landscape (Figure 199).

Figure 199. Boonville Hills east of Blairsville.

The western boundary of Boonville Hills is more rugged than Wabash Lowland. The eastern boundary is transitional to Crawford Upland's still more resistant Mississippian sandstones and limestones. Bedrock resistance to weathering and stream erosion play major roles in unglaciated landscapes.

Loess covers higher areas and is thickest in the west. This dust thins to the east, farther from the Wabash River floodplain (Figure 200). In the Midwest, loess thins with distance from floodplain sources. After westerly winds picked up dust, they eventually died and dropped this fine-grained sediment. Bedrock shows through many places in the Boonville Hills. There was extensive coal strip mining, typical of areas with Pennsylvanian coal-bearing cyclothems. The Ohio River floodplain on the Boonville Hills southern border is underlain by sandy glacial meltwater outwash. Native Americans built significant mounds on the Ohio River floodplain (Figure 201). Smaller stream floodplains are underlain by thick lake silts. These are poorly drained.

Figure 200. Loess coats uplands of Boonville Hills.

Figure 201. Native American mounds built on the Ohio River floodplain. (Angel Mounds State Historical Site)

The Boonville Hills is bounded on the north and west by the Wabash Lowland and on the east by the Crawford Upland. The Ohio River forms the southern boundary (Figure 2).

Lincoln State Park

This 2,026-acre park was once occupied by Abraham Lincoln's family from 1816 to 1830. Born in 1809 in Kentucky, Lincoln and his family moved to southern Indiana when he was seven years old. A property dispute prompted the move from a slave state to a free state where Lincoln's father, Thomas Lincoln, felt more comfortable. He lost his land holdings in Kentucky and had to start over (Figure 202).

Figure 202. Lincoln State Park occupies land where Lincoln developed as a young person.

Considerable work by the CCC built a lake, trails, campgrounds, and planted trees. **Sarah Lincoln Woods Nature Preserve** has oak woodlands on dry, sandy soil with scattered prairie. The park, established in 1932, possesses the Abraham Lincoln Bicentennial Plaza and Lincoln State Park Amphitheater.

The **Nancy Hanks Memorial** is located in the nearby Abraham Lincoln Boyhood Home and National Museum.

Weber Lake is an example of a Pennsylvanian strip coal mine reclamation project. Surface mining of coal created an environmental disaster due to acid mine drainage. Pyrite in the rock waste oxidizes and makes the water excessively acidic. The project remedied this with various methods to clear the water.

Example nature preserves in Boonville Hills

Wesselman Woods Nature Preserve has one of the largest old growth urban forests in cities of over 100,000 population. This virgin bottomland hardwood forest in Evansville has 125 trees per acre. Some are over 150 feet tall and more than 300 years old. Four of the state's largest trees are here. A tulip poplar tree is a former state champion and the tallest of its kind in the state. The preserve is a National Natural Landmark and a State Nature Preserve. It is owned by the City of Evansville.

Eagle Slough Natural Area is a wetland and mature bottomland forest domain. More than 160 species of birds have been reported. The forest has some of the largest bald cypress trees in Indiana. Bald cypress are deciduous conifers with reddish-brown bark. They are related to California redwoods.

Patoka River National Wildlife Refuge has a bottomland forest of giant cane. The state-endangered swamp rabbit lives here, along with 400 animal species. Birds account for 250 species in these wetlands along the Patoka River.

Angel Mounds State Historical Site is a fantastic Native American Mississippian culture center on the Ohio River (Figure 201). It is second in size only to Cahokia Mounds in Illinois. The Mississippian culture developed agriculture, utilizing corn as a food source. The culture ranged from 800 to 1600 A.D. (not to be confused with Mississippian Period rocks). Up to 1,000 people lived here until leaving about 1450 A.D. Although the site is in the Wabash Valley Seismic Zone, it is not known if earthquakes or other factors were involved to force people to abandon the area. Two and one-half million artifacts have been uncovered at the site. For a detailed discussion of Angel Mounds see Counts, et al. (2014).

4c. Martinsville Hills

Mississippian bedrock landscape in the Martinsville Hills was highly modified by glacial activity prior to the Wisconsinan glaciation. However, it is similar to other regions in southern Indiana with exposed bedrock that lack glacial effects. The ancient glaciations created a more subdued appearance than unglaciated areas. Pre-Wisconsinan glacial deposits do not completely cover the bedrock landscape, but instead modify it. This modification tends to smear the southern boundary with the Crawford Upland, Mitchell Plateau, and Norman Upland. Thinner pre-Wisconsinan tills to the south create a transitional boundary. South of the Martinsville Hills, unglaciated bedrock exerts more of a control on the landscape. The eastern and western ends of the Martinsville Hills are more rugged than in the middle (Figure 209). Pennsylvanian outcrops are present in the western section while the rest of the area is underlain by Mississippian strata.

Martinsville Hills are more rugged than the Tipton Till Plain and New Castle Till Plains to the north, as well as the pre-Wisconsin till plains of the Wabash Lowland to the west. To the south, the boundary is close to the edge of pre-Wisconsinan glacial deposits before entering unglaciated Crawford Upland, Mitchell Plateau, and Norman Upland where topography is strongly controlled by bedrock type. For example, topography developed on Mississippian limestones is distinctly more rugged with obvious karst sinkholes south of the pre-Wisconsin glacial boundary in the Mitchell Plateau than the same strata in the Martinsville Hills.

Martinsville Hills is bounded on the north by Tipton Till Plain and New Castle Till Plains and Drainageways. The western boundary is the Wabash Lowland. The southern boundary lies against Crawford Upland, Mitchell Plateau, and Norman Upland (Figure 2).

<u>There are no state parks in Martinsville Hills.</u>

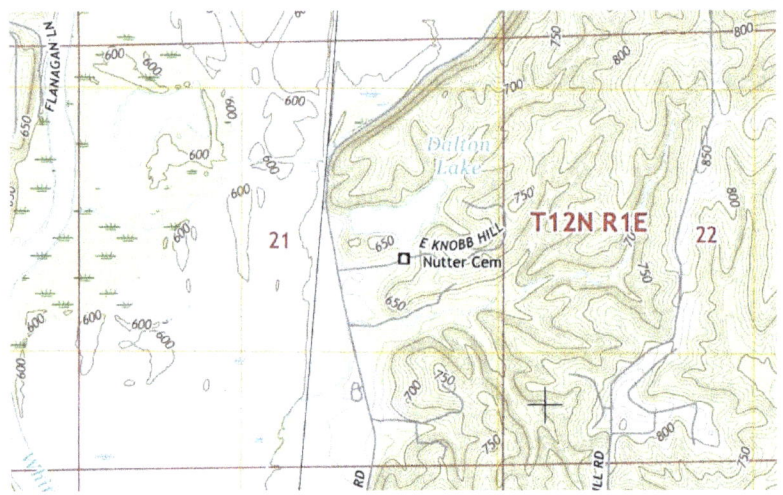

Figure 203. The eastern end of the Martinsville Hills displays the White River's wide floodplain which was enlarged by Wisconsinan glacial meltwaters passing through the drainage system. (U.S. Geological Survey Martinsville Quadrangle, (2016))

Example nature preserves in Martinsville Hills

Cataract Falls in Liebere State Recreation Area

This area features rolling hills, upland forest, and Mississippian limestone outcrops. Differences in resistance to stream erosion by layers within the Ste. Genevieve Limestone exposed along Mills Creek control topography of the stream bed. Slight differences in bedrock resistance produce small cascades. Stronger contrasts in resistivity to erosion produces significant drops to form waterfalls. Dating from Illinoian glacial times, Mill's Creek excavated through glacial lake sediments capping the limestone strata (Indiana Geological and Water Survey (IGWS), 2018). Cascades Falls waterfalls are the state's largest falls by volume.

Upper Cataract Falls drops over twenty feet. This waterfall is held up by a limy sandstone within the Ste. Genevieve Limestone (Figure 204). Cataract Falls Bridge is upstream from the falls and may be the best surviving Smith truss covered bridge.

Lower Cataract Falls drops eighteen feet. It is held up by the Lost River Chert within the Ste. Genevieve Limestone (Figure 205).

Eight feet below the top of Lower Falls the stream runs on St. Louis Limestone (Figure 206).

Figure 204. Upper Cataract Falls, Mill Creek. Mississippian Ste. Genevieve Limestone contains a limy sandstone that is more resistant to erosion than surrounding rocks. This serves as caprock for the waterfall. (Cataract Falls State Recreation Area)

Figure 205. Lower Falls, Mill Creek. Caprock is cherty limestone called the Lost River Chert Bed. (Cataract Falls State Recreation Area)

Figure 206. Panorama of Lower Falls of Cataract Falls on Mill Creek. Below the falls the stream runs on St. Louis Limestone. (Cataract Falls State Recreation Area)

Cagles Mill Lake State Recreational Area

During the Mississippian Period, Indiana was inundated by a shallow sea depositing limestones, sandstones, and shales. At the close of the Mississippian the sea retreated from the state as sea levels lowered. For several million years, Indiana stood above sea level and experienced significant stream erosion (Droste & Keller, 1989).

The sea returned at the start of the Pennsylvanian Period. The first sediments deposited in Indiana were sands brought by surface streams to the sea from the east. This produced the oldest Pennsylvanian stratum in Indiana, the Mansfield Formation.

Glaciation in the southern hemisphere caused sea level to rise and fall many times during the Pennsylvanian. This resulted in a shifting coastline that repeatedly advanced and retreated. Deltas, sluggish sediment-laden streams, swamps, and many microenvironments changed from place to place. This gave rise to sequences of environmentally-controlled strata called cyclothems. Although the rising and falling of sea level was cyclic, that does not mean every microenvironment of every cycle is preserved. Since part of the falling sea level of a cycle involved exposure of previously deposited sediments, stream erosion often removed some of these layers. That accounts for "missing" environments in the Pennsylvanian strata.

How do we "read" ancient environments?
1. Limestones almost universally represent marine environments.
2. Shales can be of marine or nonmarine origin.
3. Sandstones usually indicate stream, nearshore, or beach action. Sandstones in Pennsylvanian cyclothems often result from streams cutting a channel and are frequently part of a delta complex. Examples are seen in the walls of Cagles Mill Lake Spillway.

Cagles Mill Lake Spillway

Cagles Mill Lake spillway provides an amazing array of exposed strata not seen in the rest of the area. The following interpretations follow IGWS (2018) and Langenheim et al. (1966). This huge outcrop displays Pennsylvanian Mansfield and Brazil Formations. A large stream channel was carved into the Mansfield sandstone. The channel was filled with coal, shale, and sandstone (Figures 207-211). These deposits represent the dynamic and ever-changing environments in the lowlands near the Pennsylvanian shoreline. Delta development due to shifting streams depositing sediment made a variety of location conditions. Swamp vegetation, once buried under other sediments, converted to coal, a significant energy source for Indiana. At the far east end of the spillway outcrop, Pennsylvanian rocks are truncated by an unconformity (eroded land surface) with a time gap of three hundred million years. They are topped by pre-Illinoian and Illinoian sediments, Sangamon interglacial soil, and Wisconsinan loess.

Figure 207. The Mansfield sandstone (left) was eroded by a large stream. This channel was filled with Brazil Formation (right and topped the Mansfield in the center left). (Panorama of Cagles Mill Spillway looking east)

Figure 208. Pennsylvanian Mansfield thick-bedded sandstone (left) was cut into by a large river channel which was later filled by coal, black shale grading to gray shaly sandstone and topped by thin-bedded sandstone with tree root fossils (*Stigmaria*). The channel fill is part of the Brazil Formation. The Mansfield sandstone was itself also part of an older channel filling. (Cagles Mill Spillway, north wall)

Figure 209. Center of the Brazil Formation channel fill. The bottom of the channel is black coal (thicker in middle, three feet), followed by black shale, gray shale becoming more sandy, and topped with thin-bedded sandstone. Weak rocks fall to make a talus pile at the base of the slope. (Cagles Mill Spillway, north wall)

Figure 210. The south wall of Cagles Mill Spillway exposes the center of the Brazil Formation channel fill, partly covered by talus. Thin-bedded sandstone caps gray shale. At the west end of the spillway are tree roots (*Stigmaria*) and tree trunks (*Lepidodendron,* an extinct scale tree).

Figure 211. Pennsylvanian cyclothems represent continually changing delta environments. (Cagles Mill Spillway, north wall)

4d. Crawford Upland

Glaciers did not touch the Crawford Upland. That does not mean that all its boundaries are sharp compared to its neighboring areas which bear the marks of pre-Wisconsinan glaciation. The northern boundary is unclear due to the work of those ancient glaciers (see discussion about Martinsville Hills).

The upland's western boundary is gradational. Glacially subdued landscapes and weak Pennsylvanian shale-sandstone-coal-limestone beds fade into Crawford Upland. The upland is more rugged and is developed on Mississippian sandstone-limestone strata.

Strata dip (tilt) southwest off the eroded Cincinnati Arch. This accounts for progression of sedimentary rock ages from younger on the southwest to older rocks northeast. The weaker Pennsylvanian strata form an unspectacular east-facing *escarpment* where the Wabash Lowland and Boonville Hills meet the Crawford Upland. This transitional boundary is different south of the Ohio River. There the Pennsylvanian sandstones are better cemented. They form the Pottsville Escarpment in Kentucky. This illustrates how important the cementation of sandstones is to topography. Rocks exposed further east in the Crawford Upland display various rugged topographic elements depending on the amount of sandstone or limestone present. Soluble Mississippian limestones east of the upland permit groundwater solution to form sinkholes and caves. Sandstones form ridges with steep slopes. Flat-topped uplands are common. The slight dip of strata and deep stream incision produce ledges on these uplands.

The eastern boundary of the Crawford Upland is the sharp Springville Escarpment that can be traced from the Ohio River north to where it disappears under glacial sediments near Spencer. An older term is the Chester Escarpment, based on the Mississippian Chesterian strata that hold up this landscape feature. These more resistant sandstones and limestones are responsible for this topographic break with weaker Mississippian limestones on the east side (Figures 212-216). The break is not a straight line. Stream erosion breaks up the escarpment as the front wears away. This leaves isolated steep-sided hills surrounded by low-lying limestone pock-marked with sinkholes.

Figure 212. Springville Escarpment (Chester Escarpment) separates the rugged Crawford Upland capped by sandstone and limestone to the west from the more subdued Mitchell Plateau of soluble limestones (foreground). All strata are Mississippian Period rocks. View looks west at the tree-covered edge of the cuesta. (IN 64 east of Milltown)

The Dripping Springs Escarpment in Kentucky is equivalent to the Springville Escarpment in Indiana. Hills or ridges with escarpments on the leading edge of gently dipping resistant strata are termed *cuestas*.

Figure 213. Resistant Mississippian rocks exposed along the Springville Escarpment. (IN 64 just east of Milltown)

Figure 214. Mississippian sandstone and limestone strata hold up the rugged topography along the Springville Escarpment. (US 64 quarry east of Milltown)

Figure 215. Resistant Mississippian strata of Crawford Upland near east edge of Springville Escarpment. (I 64 near 98.5 mile)

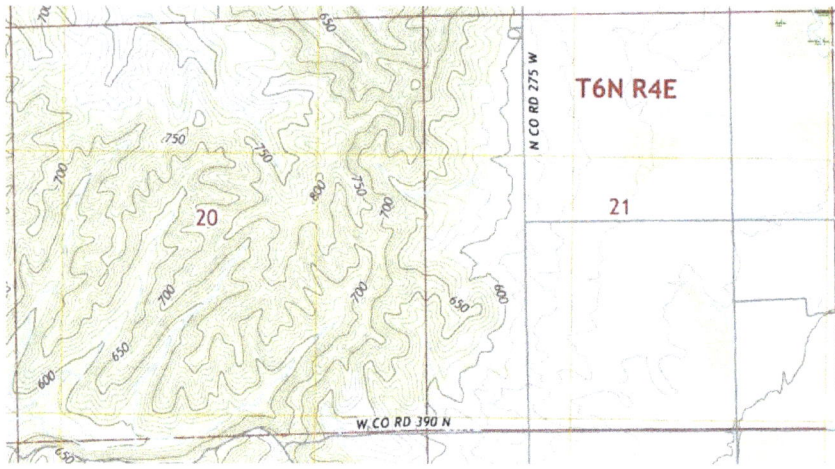

Figure 216. Springville Escarpment separates the rugged Crawford Upland on the west and low relief of the Mitchell Plateau to the east. Compare with Figure 212. (U.S. Geological Survey Brownstown Quadrangle, (2016))

Elevations of the Crawford Upland range from 450 on the west to 900 feet on the east. The Crawford Upland has experienced deep incision by the White, East Fork of the White, and Ohio Rivers. Their ridges and valleys form a rugged landscape with 200 to 300 feet of local relief.

The Crawford Upland is bounded on the north by the Martinsville Hills, on the east by the Mitchell Plateau, and on the west by the Wabash Lowland and Boonville Hills (Figure 2).

O'Bannon Woods State Park

Established in 2004, O'Bannon Woods' 2,300 acres is part of the much larger **Harrison-Crawford State Forest**. The region borders on the Ohio River and displays ridges, knobs, caves, sinkholes, scenic 200-hundred-foot-high river bluffs and steep gorges carved in Mississippian strata (Figures 217). Lower elevations expose limestones. The ridges are capped by sandstone and limestone. The northern boundary is the Blue River, said to be the most beautiful river in Indiana (Figures 218 and 219). The Blue River in this vicinity runs on limestone. Majestic vistas can be seen on the Ohio (Figures 220 and 221).

Figure 217. Ohio River bluff with Mississippian limestone boulders which have fallen from the cliff by weathering. (O'Bannon Woods State Park)

Geology of Indiana State Parks

Figure 218. Mississippian limestones exposed in the Blue River valley. (Intersection of IN 62 and Harrison Spring Road)

Figure 219. Blue River near O'Bannon Woods State Park. The river runs on Mississippian limestone. (Intersection of IN 62 and Harrison Spring Road)

Figure 220. Ohio River bluffs across the river in Kentucky tower over the majestic river that drains an area where ten percent of the United States population lives. Mississippian bedrock at lower elevations is limestone. Higher elevations are sandstone and limestone. (O'Bannon Woods State Park)

Figure 221. Ohio River meander. Bedrock is Mississippian Period strata. Lower elevations are limestones. Ridge caps are sandstone and limestone. (IN 62 overlook near Leavenworth)

O'Bannon Woods was the location of one of the few African American CCC units. Within the park are nature preserves:

Post-Oak-Cedar Nature Preserve displays slopes of limestone and sandstone.

Mouth of the Blue River Nature Preserve consists of several knobs, deep ravines and steep bluffs.

Wyandotte Caves (Big and Little Wyandotte, and Easter Pit) (National Natural Landmark) are open only from Memorial Day to Labor Day on weekends (see website to verify dates and times). The caves are dissolved out of Mississippian limestone. Native Americans entered the cave 4,000 years ago to find chert, epsomite (for medicinal purposes), and aragonite (similar to calcite).

Example nature preserves in Crawford Upland

Tank Spring Nature Preserve: A spring emerges from Mississippian limestone at the base of a sandstone cliff. The water was used in the past for steam-powered train locomotives. Productive soils of moist upland and bottomland forests are in the preserve. Hunting is permitted in season. Be careful.

Bluffs of Beaver Bend Nature Preserve: On a meander bend of the White River is a mile-long 150-foot-high sheer sandstone bluff. Native Americans used the bluffs as shelter 750-1,500 years ago and left three six-foot-tall mussel shell piles. No trail. Cliffs and rock shelters of Mansfield sandstone are exposed along the length of the bluff.

Patoka Lake Hiking Area features the boundary between Mississippian Waltersburg sandstone and overlying Pennsylvanian Mansfield sandstone. Totem Rock is of Mansfield sandstone. The sandstone displays cross-bedding (angular bedding in a horizontal stratum produced when the original sand was deposited on slopes such as ripples/dunes in streams), honeycomb weathering, and *liesegang rings* (red/brown colored bands or rings resulting from chemical precipitates of iron in the rock; the iron is insoluble and stands in relief).

Hemlock Cliffs Special Place, Hoosier National Forest, Patoka Unit is a remote box canyon exposing Mississippian Tar Springs Formation sandstone outcrops with an acclaimed waterfall. Waterfalls in the place are tall and seasonal, which means you may see nothing more than a trickle of water or high flow, depending on precipitation. There are overhangs, rock shelters, cliffs, ravines, and honeycomb weathering. Underlying the sandstone is the Mississippian Glen Dean Limestone with its springs and small caves. Native American occupation was apparently over 10,000 years ago. At the head of the box canyon is a large, semicircular rock shelter, which was likely used as cover and defense by early Native Americans. Parts of the trail are steep and slippery when wet, so use caution. The trail is a one mile loop.

Buzzard Roost Recreation Area, Hoosier National Forest, Tell City Unit features significant Ohio River sandstone bluffs. Buzzard Roost Trail is an 0.8-mile hiking trail that begins on the bluff overlooking the Ohio River and descends to the river. There are several sandstone outcrops, bluffs, waterfalls, and views of the Ohio River. At a lookout, notice the steep cutbank of the river on the outside of a meander. Across the river on the Kentucky side is the point bar of the floodplain and bluffs.

Clover Lick Barrens/Clover Lick Special Area, Hoosier National Forest, Tell City Unit near the Ohio River has thin soil developed on rocky limestone bedrock.

German Ridge Recreation Area, Hoosier National Forest, Tell City Unit offers a trail with sandstone bluffs.

Green's Bluff Nature Preserve is considered one of the most scenic in Indiana. There are steep and rugged sandstone cliffs on Raccoon Creek, limestone bluffs and caves, and a floodplain forest. There are two segments of the reserve with separate trails.

Porter West Preserve hosts sandstone outcrops and limestone karst (sinkholes and spring).

Harrison Spring (National Natural Landmark) is the largest spring in Indiana. It is private property and is not open to the public.

Marengo Cave (National Natural Landmark) has two commercial cave tours (The Crystal Palace and Dripstone Trail). Cave temperature is 52°F. These tours involve easy walking but are not handicap accessible and involve climbing stairs.

Jug Rock Nature Preserve is a free-standing rock pillar ("tea-table") made by differential erosion and weathering of Pennsylvanian Mansfield sandstone (Figure 222). The pillar is 20 feet wide and 60 feet high. The rock displays cross-bedding (Figure 223), indicating the flow direction of braided streams carrying sand across a delta about 300 million years ago. Oxidized iron coloration represents groundwater deposition of iron in rock. Once exposed to air, iron oxidizes and resists weathering of the rock.

The pillar separated from a sandstone cliff by weathering and erosion along fractures in the bedrock strata. The resistant caprock limits destruction of the pillar, although the layer below the cap is softer and recessed (Figure 224). Jug Rock is claimed as the largest freestanding rock formation east of the Mississippi River.

To reach Jug Rock from Highway 50 west of Shoals, go past the Shoals Overlook rest park. About 200 yards downhill from the park is a small pull-off from the highway. The parking spot is easy to miss. Don't turn onto an adjacent road. Park only in the pull-off and follow the rugged trail into the woods to the rock. Nearby House Rock has a similar origin.

Figure 222. Jug Rock, a pedestal isolated from the rest of rock layers by erosion. (Jug Rock Nature Preserve)

Geology of Indiana State Parks

Figure 223. Cross-bedding indicates current flow from left to right. Recessed strata sand grains are less well cemented than layers above and below and finer grain size.

Figure 224. Caprock of well cemented grains over a layer of weakly cemented smaller grains. Honeycomb weathering is due to washing out of weakly cemented areas, leaving depressions.

Yellow Birch Ravine Nature Preserve is in Hoosier National Forest near Taswell. It features waterfalls, especially Bowl Falls and Double Falls, plus Ravine Arch.

Bluespring Caverns is developed in Mississippian Salem Limestone. Boat and walking tours are available. The above ground walk includes what is described as Indiana's largest sinkhole. There is an entry fee.

Indiana Caverns is the longest cave in Indiana. Most of the cave is in St. Louis Limestone. The upper level ceiling is in Ste. Genevieve Limestone. A large number of Pleistocene mammals fell to their deaths in a fissure at Indiana Caverns. A boat tour is available. There is an entry fee.

Patoka Lake within the Newton-Stewart State Recreation Area features a trail exposing Mississippian sandstones displaying cross-bedding emphasized by iron oxide staining. Rocks are weathered into overhanging bluffs, arches, and hollows. Totem Rock is a block of sandstone which has separated from the rest of the rock layer along vertical joints and slid downhill. Various formation names have been applied to this sandstone layer. If interested in the geological details of formation names, see Indiana Geological and Water Survey's Indiana Geologic Names Information System.

4e. Mitchell Plateau

This Mississippian limestone *plateau* is unglaciated except for the extreme northern area where it transitions to the Martinsville Hills (see that discussion). There is significant karst development, especially in the eastern section (Figure 225). Mississippian limestones dominate the bedrock (Figure 226) and display abundant sinkholes and caves.

Figure 225. Panorama of rolling hills produced by innumerable sinkholes. (South of Spring Mill State Park)

The western edge is at the base of the Springville Escarpment (Figure 227). The plateau is deeply incised by streams into thick Mississippian limestones (Figure 228). The eastern boundary is the eastern edge of Mississippian limestone bedrock. North of the East Fork of the White River are a number of deeply entrenched large streams. Limestone ledges form steep slopes. South of the river, the eastern plateau area consists of rolling hills covered by a weathering residue of red clay, which results from dissolving of limestone and oxidation of iron in the rocks. Such residues are common in karst areas of the Midwest, e.g., Missouri, Kentucky, Tennessee. They are often termed *terra rossa*. Caves, sinkholes, karst springs, and sinking streams are abundant. Areas densely populated with sinkholes have little visible surface drainage.

Figure 226. Mississippian limestone with cross-bedding (current flow was to left at top and to right at bottom). Honeycomb weathering marks where water dissolves weakly cemented spots. (IN 37 south of Bloomington)

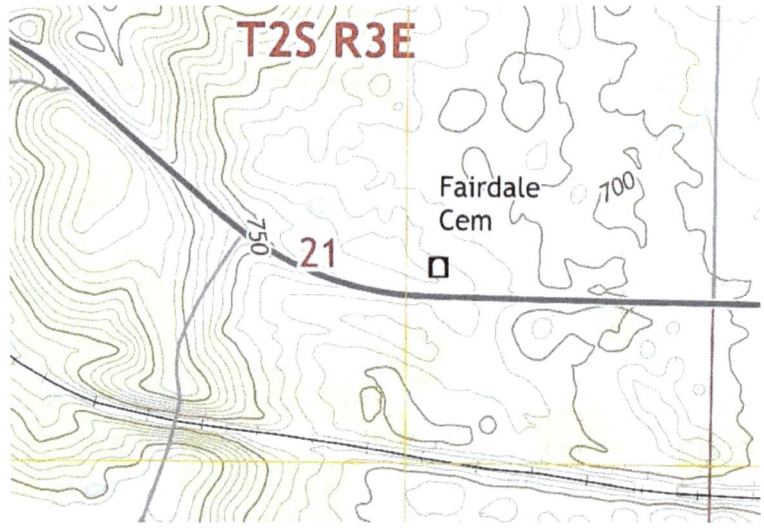

Figure 227. Rugged Springdale escarpment separates high relief Crawford Upland (left) from Mitchell Plateau of soluble limestones with sinkholes (hachured contours) and sinking streams. (U.S. Geological Survey DePauw Quadrangle, (2016))

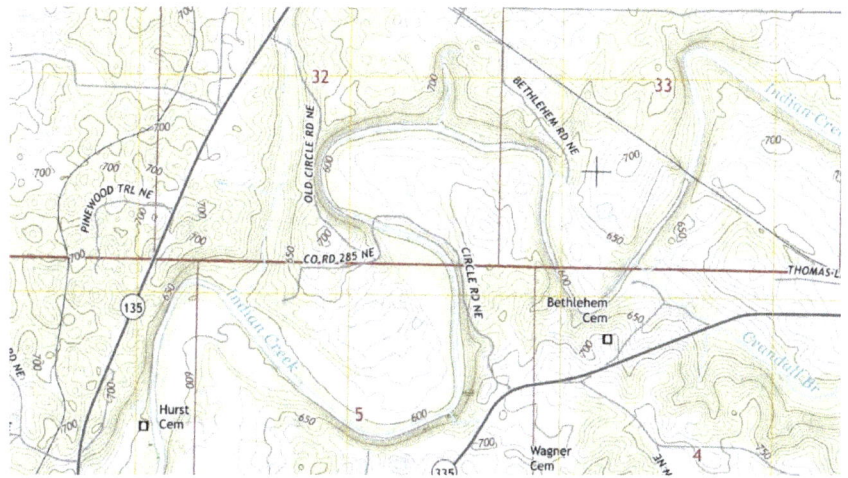

Figure 228. Deeply incised stream on Mitchell Plateau. Note intense karst development (sinkholes) to the west. (U.S. Geological Survey Crandall Quadrangle, (2016))

The Mitchell Plateau contains a classic karst area made well known by Malott (1922; 1932; 1952). The western edge where the limestones lie beneath younger Mississippian sandstones and limestones is easily recognized by the rugged character of the Crawford Upland to the west (Figure 227). However, this boundary (Springdale Escarpment) is not a straight line. Instead, it is broken by stream erosion into an irregular patchwork of isolated hills capped by resistant upland strata surrounded by low-lying limestones riddled with sinkholes.

The Mitchell Plateau is bounded on the north by the Martinsville Hills, on the east by the Norman Upland, and on the west by the Crawford Upland (Figure 2).

McCormick's Creek State Park

Long before glaciers traversed the McCormick's Creek area, Mississippian limestones were being deposited in a warm, shallow sea. Myriads of invertebrates and microscopic organisms found the environment much to their liking. Their shells accumulated to form hundreds of vertical feet of limy sediment. Eventually these thick layers of lime mud and fossil fragments were buried under more sediments. Pressure compacted the materials and buried seawater cemented the grains together with calcite to make solid limestone. Hundreds of millions of years passed after the seas covered Indiana. Tectonic uplift of the land drove away the seas and allowed stream erosion to wear through layer upon layer of strata. Eventually the limestones were exposed to rainwater, which moved underground along fractures and bedding surfaces. The result is a karst landscape of caves and sinkholes.

A million years or so before the present, the first of several waves of glacial ice invaded the already well-eroded landscape. Eventually the preglacial streams were blocked, and McCormick's Creek was a new avenue to carry away Illinoian glacial meltwater.

McCormick's Creek State Park is the first of Indiana's state parks. The park's 1,924 acres was established in 1916 and surrounds a mile-long canyon carved in Mississippian limestones that display a scenic waterfall. This is a classic karst landscape with abundant sinkholes, intermittent streams flowing over limestone, and a cave with arches. A resistant layer in the St. Louis Limestone caps a waterfall underlain by weaker strata. Having strata of contrasting resistance is important in the development of a waterfall. The top of the canyon walls is held up by Ste. Genevieve

Limestone. McCormick's Creek has incised the limestones on its way to the West Fork of the White River (Figures 229-231).

Figure 229. Panorama of McCormick's Creek valley. Top of bluffs are Ste. Genevieve Limestone (compact, fine-grained, smooth). Most of the canyon walls and caprock of the waterfall are St. Louis Limestone (fine-grained, thin bedded with thin layers of shale, dolostone, and chert). Downstream and visible in an old quarry is the Salem Limestone (thick bedded, granular texture), which was used as a building stone. (McCormick's Creek State Park)

The Statehouse Quarry in Salem Limestone operated from 1878 to 1880, providing foundation and basement stone for the state capital in Indianapolis. The Salem displays cross-bedding, indicating current motion working the shelly material when the sediment was deposited in the shallow seas of Mississippian time. The karst landscape of caves and sinkholes developed on the soluble limestones are present throughout the park. Most of the sinkholes are in the Ste. Genevieve and St. Louis Limestones.

Geology of Indiana State Parks

Figure 230. Waterfall lip is a resistant layer in the St. Louis Limestone. Less resistant layers below wear away and collapse as the waterfall retreats upstream. (McCormick's Creek State Park)

Illinoian glacier ice covered the area and ended its trek a short distance from the park. The region's drainage system was filled with outwash sediment. Meltwater shifted the region's drainage direction from west to northwest. This is evident by the direction creeks move through the park. It is thought that erosion of the canyon was more rapid during wet glacial times of Wisconsinan glaciation.

Figure 231. Weak shales of the St. Louis Limestone collapse from weathering and erosion, widening the valley.

Wolf Cave Nature Preserve in the park displays a small cave and Twin Bridges arches, a former cave with both ends exposed by surface erosion cutting into the rock from two sides.

Spring Mill State Park

This 1,358-acre park was established in 1927. Spring Mill is karst country plus! There are caves, sinkholes, sinking streams, and karst springs galore (Figure 232). Karst springs provided water for gristmill and sawmill work in the 1800s. The CCC built trails, fences, picnic areas, shelters, and restored buildings. There is a memorial to Virgil I. Grissom, the second American in space aboard the Mercury project Liberty Bell 7.

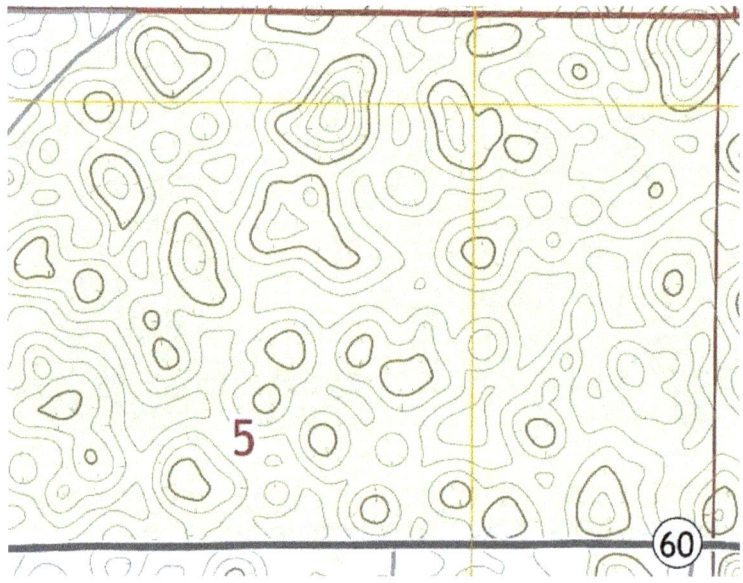

Figure 232. Closed depressions (dashes point to center) are sinkholes where surface water drains directly into caves in Spring Mill State Park. The lack of surface streams is characteristic of intense karst development. (U.S. Geological Survey Mitchell Quadrangle, (2016))

At Twin Caves, you can visit a *karst window* where the roof of a cave collapsed (Figures 233-236). This exposes a cave stream as it enters one end of the cave and exits the other end. Within the cave, blind cave fish may be seen on the boat tour of Upper Twin Cave during the summer. A fee is required for the tour.

The two Mississippian limestone formations in the park, the Salem and St. Louis, sport many cave passages. The Salem can be seen at cave entrances to Hamer and Donaldson Caves. The St. Louis Limestone contains caves and most of the sinkholes.

Figure 233. Upstream cave entrance and exposed cave stream of Twins Caves karst window. (Spring Mill State Park)

Figure 234. Downstream cave entrance of Twin Caves karst window. (Spring Mill State Park)

Figure 235. Panorama of Twin Caves karst window. Upstream cave entrance is far left next to the stairs and downstream cave entrance is far right (diagonal log on the side). (Spring Mill State Park)

Figure 236. Downstream half of the karst window (cave is dark hole to right). (Spring Mill State Park)

Cave River Valley Natural Area is a restricted part of Spring Mill State Park. Permit or guided tour reservations are required for entry. Be aware that hunting is permitted in season. A trail may not be well marked.

The **Mitchell Karst Plains Nature Preserve** is part of Spring Mill State Park. This is a classic karst sinkhole plain by the incised valley of Mill Creek. Sinkholes in various stages of development can be seen. Some drain after a storm while others are plugged with sediment and vegetation to form temporary ponds. This may be the largest area of undisturbed sinkhole plain in an Indiana forest.

Donaldson Cave/Woods Nature Preserve (National Natural Landmark) is in Spring Mill State Park and is considered the most picturesque location in Indiana with *mesophytic forest* between beech-maple and oak-hickory trees. This is one of the most impressive stands of the original forest remaining in Indiana. Some trees are over 300 years old. Much of the water enters sinkholes with little stream runoff. The caves may be closed due to White Nose Syndrome danger to bats. Permission must be obtained, and training is required if the caves are open.

Example nature preserves in Mitchell Plateau

Wayne Woods is considered the smallest nature preserve in Indiana. Several sinkholes are in the woods. There are over 1,000 native species of plants and animals recorded.

Leonard Spring Nature Park consists of two karst springs. Shirley Spring Cave is a breathing cave, due to warm moist air that condenses to steam on contact with cold winter air. The spring drains into wetlands.

Cedar Bluffs Nature Preserve contains a 75-foot bluff of Mississippian limestone, the lower slope is Harrodsburg Limestone and Salem Limestone caps the ridge top. The valley is cut by a tributary of Clear Creek. The preserve lies on the Springville Escarpment which marks the boundary of the Mitchell Plateau and Crawford Upland.

The Cedars Preserve is a remote forest that exposes Mississippian limestone with sinkholes and an abandoned limestone quarry. The preserve was closed at the time of this writing.

Orangeville Rise of the Lost River Nature Preserve (National Natural Landmark) is a three-acre plot and has the second largest spring in Indiana. The underground watershed drains from 30 square miles of karst and scenic hills. Sinkholes in the region are abundant (up to 1,000 per square mile) (Figures 237 and 238). *Losing* or *sinking* streams are common throughout the area (Figure 239). **Lost River** is the largest sinking stream in the United States (Figure 240). It sinks into swallow holes on a farm upstream from the preserve. The dry creek bed downstream carries water only during heavy rain. Artesian water rises to the surface off preserve property and also on the preserve from a cave (Figures 241-244) into a pit 220 feet across at the base of a low cliff of Mississippian Ste. Genevieve Limestone. The underground water system contains over 24 cave species (including blind cavefish). The entire Lost River drainage watershed area is over 350 square miles in five counties.

Figure 237. Sinkhole pond south of Spring Mill State Park.

Figure 238. Sinkhole pond during drought. (South of Spring Mill State Park)

Geology of Indiana State Parks

Figure 239. Dry creek bed of a losing stream. (East of Orangeville)

Figure 240. Lost River before it sinks underground. (South of Spring Mill State Park)

Geology of Indiana State Parks

Figure 241. Orangeville Rise and stream issuing from the spring. (Orangeville Rise of the Lost River Nature Preserve)

Figure 242. Panorama of Orangeville Rise and stream. The spring represents part of the flow of the Lost River system as it reaches the surface. (Orangeville Rise of the Lost River Nature Preserve)

Figure 243. Green color indicates high biological activity.

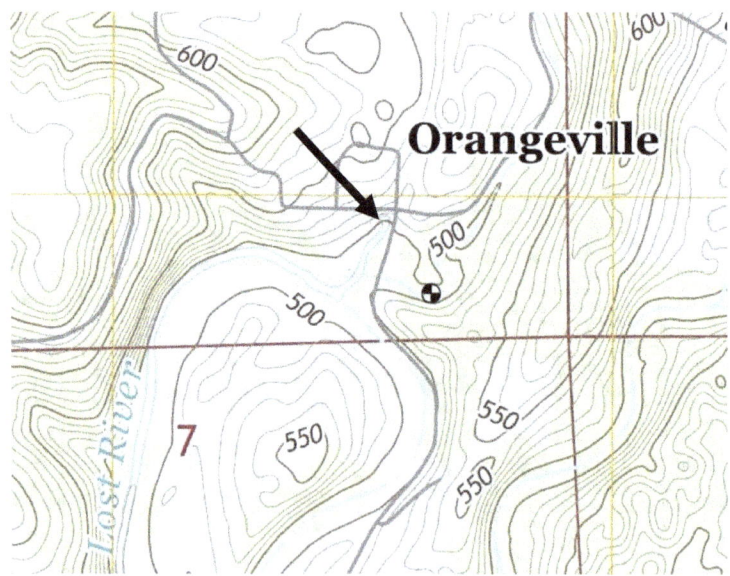

Figure 244. Orangeville Rise spring (arrow) feeds Lost River, intermittent upstream from the spring. Note sinkholes to north. (U.S. Geological Survey Georgia Quadrangle, (2016)**)**

Wesley Chapel Gulf, Hoosier National Forest, Lost River Unit (National Natural Landmark) is a 1,075 by 350-foot sinkhole that formed when a cave roof collapsed and exposed Lost River flowing underground. Lost River flows five miles underground before reaching the karst window. A 125-foot-wide rise (Boiling Spring), marks where Lost River enters the sinkhole before disappearing at the other end. Water can rise in the gulf from 20 to 50 feet, depending on precipitation. Mississippian limestone bedrock is exposed in the walls. Muddy conditions can be dangerous. Orangeville Rise is downstream from Wesley Chapel Gulf.

Twin Creek Valley Nature Preserve and Henderson Park has limestone glades and outcrops, waterfalls, and wet caves. Call or check in with the police department at Salem City Hall before entry.

Big Spring Nature Preserve features a cave spring. The minimum flow rate is 450,000 gallons per minute. The rise pit is 15 feet across and six or eight feet deep. The site is in high-quality old-growth forest. Avoid this preserve when activities are going on at Big Springs Church since the trail begins behind the church.

Hayswood Nature Reserve features scenic rock outcrops of Mississippian limestone that overlook Big Indian Creek 300 feet below. The site is located on one limb of a meander of the creek.

Teeple Glade Nature Preserve is a Mississippian limestone *glade* on a steep creek exposure. Deer hunting is allowed in season, so be careful visiting the site.

Flatwoods County Park is the site of a glacial lake. Illinoian glacial ice stopped near McCormick's Creek State Park. Since glacial ice blocked stream outlets, meltwater ponded in front of the ice. Meltwater brought sand, silt and clay into the lake, creating a flat surface with up to 70 vertical feet of sediment on top of limestone bedrock. This lowland is ringed by Mississippian bedrock limestone. Glacial Lake Flatwoods eventually drained after the ice front melted back. Sinkholes and McCormick's Creek drain the area today. There is an artesian well and several springs.

4f. Norman Upland

Norman Upland is a southern Indiana region with only a partial glacial pedigree. Pre-Wisconsinan glaciation touched only the northern edge (see discussion on Martinsville Hills) and part of the eastern side of the upland. However, glaciers did little to directly influence the topography. The resistant Mississippian Period Borden Group siltstones and sandstones are responsible for this most rugged landscape in Indiana (Figures 245-248).

Figure 245. Mississippian Borden Group siltstone/sandstone. (I-64 mile 116)

Weakly resistant shales are also present, which provides some contrast in topography. The lack of strongly contrasting rock types generally yields steep but smooth slopes.

The boundary on the north is gradational into the Martinsville Hills and is close to the southern edge of pre-Wisconsinan glacial deposits. A few miles south of this glacial boundary the rugged topography and high relief of the unglaciated Norman Upland is clearly revealed. On the west the boundary is unclear but defined as the eastern edge of Mississippian limestone bedrock. The rugged landscape of the Norman Upland contrasts with the subdued shale bedrock of the Scottsburg Lowland to the east. This eastern boundary of the Upland is sharply defined at the base of the **Knobstone Escarpment.** This bold feature is the most striking physiographic feature in Indiana (Figure 246). The landscape of flat-topped ridges and deep valleys stands in contrast to areas around it. The escarpment rises 600 feet above the Ohio River and extends 150 miles north to where it is covered by Wisconsinan glacial sediments. In the subsurface, the buried escarpment extends another 100 miles to the northwest. This is an excellent example of how contrasting the landscape appeared before Wisconsinan glaciers completely buried this significant topographic feature. Pre-Wisconsinan glaciers had a tough time getting over the escarpment into the Norman Upland but one or more managed to top it. They left little trace of their success.

Figure 246. Mississippian Borden Group siltstone/sandstones hold up the landscape near the eastern edge of the Knobstone Escarpment. (I-64 mile 116)

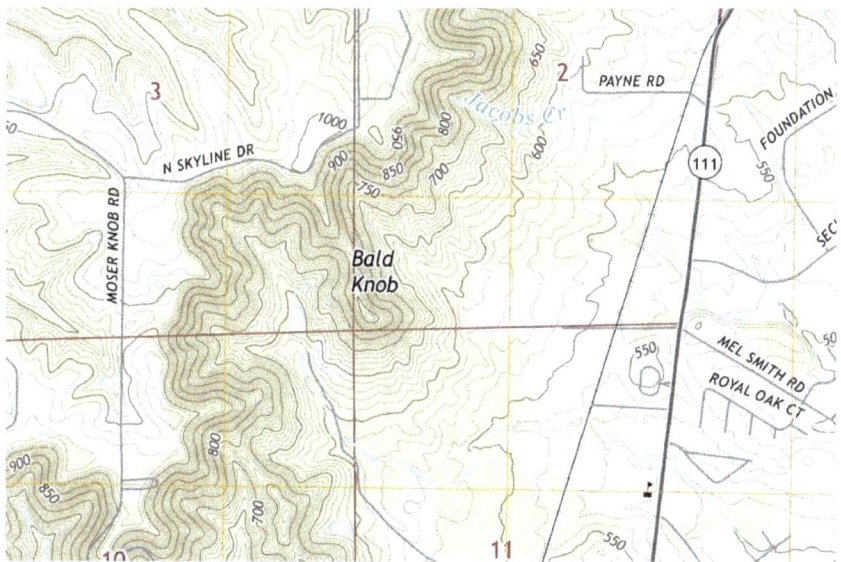

Figure 247. Large upland flats of southern Norman Upland are underlain by resistant siltstone/sandstone on west. Rugged Knobstone Escarpment is in the middle. Easily eroded Scottsburg Lowland shale on the east. Elevation drops 450 feet from west to east forming one of Indiana's sharpest physiographic boundaries. (U.S. Geological Survey New Albany Quadrangle, (2016))

The Knobstone Escarpment is held up by the Mississippian Borden Group of fine-grained sandstones and siltstones. East of the base of the escarpment are the older, weakly resistant Mississippian New Providence Shale and Mississippian/Devonian New Albany Shale. These rocks illustrate the huge impact that resistance to surface weathering and erosion have on landscapes. Northern Norman Upland was incised by pre-Wisconsinan meltwater that cut deep valleys, leaving few upland flats. The area is a maze of ridges (Figure 248). South of this region there is less incision with large upland flats (Figure 247). Remnants of the Knobstone Escarpment occupy the area south of the East Fork of the white River.

Weathering of siltstone yields non-nutritious soils. This is one reason the Norman Upland supports woodlands but not farming.

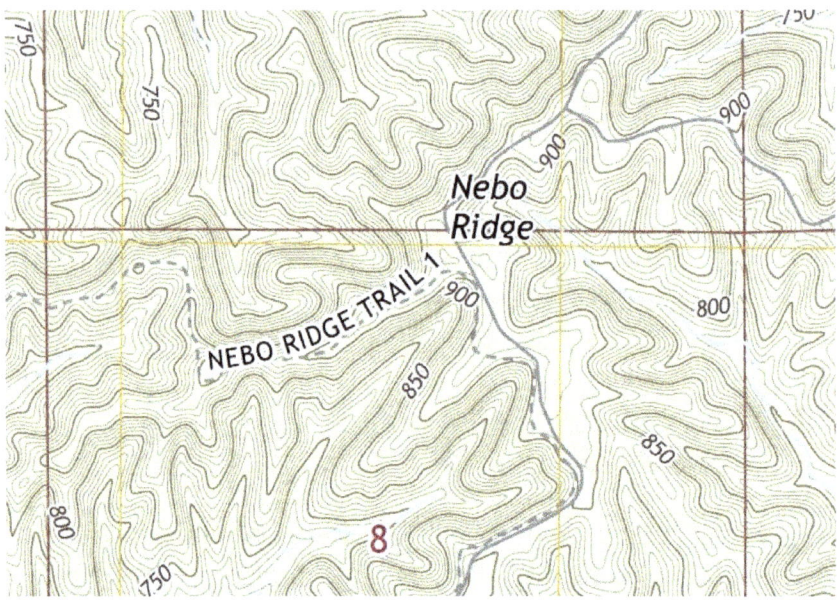

Figure 248. Maze of ridges and deep valleys in the northern section of the Norman Upland. Note the steep slopes. U.S. Geological Survey Story Quadrangle, 2016)

The Knobstone Escarpment appears as a wall from the Scottsburg Lowland. Traced north, streams dissect strata to produce "The Knobs," isolated hills rising above the lowland. Brownstown Hills is a large, isolated remnant of Norman Upland on the Scottsburg Lowland (see the Scottsburg discussion).

Norman Upland is bounded on the north by the Martinsville Hills and a narrow patch where the Knobstone Escarpment is covered by Wisconsinan till of the New Castle Till Plains and Drainageways, on the west by Mississippian limestone of the Mitchell Plateau, and on the east by the Scottsburg Lowland and Charlestown Hills (Figure 2).

Brown County State Park

This well-visited park was established in 1929 and is the largest in Indiana at 15,696 acres. The park is sometimes called Indiana's Little Smokies because of its superficial resemblance to the Great Smoky Mountains. The geology of the two areas is strikingly different but the appearance is similar. See the National Park Service website for a discussion of geology of the Appalachian Great Smoky Mountains.

The story of Brown County State Park dates to the early part of the Mississippian Period. A delta was being deposited where rivers entered the shallow sea covering much of Indiana. These sediments were washed off the Appalachian system of mountains to the east. Vast amounts of sediment spread out as sheets of sand, silt, and clay. Depending on the specific depositional environment on the delta, streams deposited particles which were later buried and transformed into shale, siltstone and sandstone by compaction and cementation of mineral precipitates in spaces between grains. The energy of a stream (its speed) determines the size of particles which water can move. Clay and silt grains are more easily moved than sand and are deposited only when currents slow significantly. Sand is eroded and deposited at higher speeds than clay or silt. Deltas are dynamic features which receive river water moving fast or slow depending on rainfall in the stream's drainage area. As rivers reach sea level, they slow down and deposit sediment. Rivers wander about a delta as river channels shift from place to place. The processes involved produce a complex mixture of environments, including marshes/swamps, river channel fills, and other deposits. The environments often change rapidly during floods.

Sea level rose and the shoreline shifted east. Sediment patterns changed as the shoreline moved. Marine conditions took over the responsibilities of sedimentation by providing huge sheets of limestone. Shells of marine invertebrates accumulated in untold numbers. Layer upon layer of various sediment patterns buried the delta's sediments throughout the rest of the Mississippian and Pennsylvanian Periods. Compaction and cementation of the delta's sediments turned them into solid rock.

At some point, tectonic stresses from the Appalachian region buckled the strata upward. This exposed the layers to stream erosion. Eventually, younger rocks were stripped away, leaving two Mississippian formations as the mainstays of exposed bedrock that hundreds of millions of years later would be eroded by Pleistocene glacial meltwaters. The older Spickert Knob Formation shales were more easily eroded than sandstone of the younger Edwardsville Formation. Thus, the Edwardsville caps resistant knobby peaks of the Knobstone Escarpment. The highest point is Weed Patch Hill 1,058 feet above sea level. This hill is further capped by the next younger Sanders Group limestone which contains *geodes*. Geodes are elliptical precipitates made by groundwater moving through bedrock. They typically have a partial hollow center with crystals growing from the walls toward the interior. Contrasts in elevation between the shales and sandstone yield up to 375 feet of relief in Brown County State Park.

There is little record of what Indiana was like after the Pennsylvanian Period. Erosion dominated until the Pleistocene glaciers began to rumble south, covering most of the state. Pre-Illinoian, Illinoian, and Wisconsinan glaciers came close to the Brown County State Park area, but they never broke over the landscape. Illinoian ice came closest but meltwater from this ice sheet did carry huge amounts of meltwater down Salt Creek. These masses of sand and gravel filled some of the valley. After the ancient glacier melted away and the ice front retreated from the area, less water rushed through the lowlands. Salt Creek began to cut into the alluvium. This produced flat terraces on the sides of the valley.

The relatively uniform erodibility of exposed siltstones and sandstones in the park yields a somewhat smooth rolling topography. Be certain to visit the overlooks throughout the park to gain a sense of this symmetrical arrangement of hills and valleys (Figures 249-253). Sandstone holds up hilltops and fine-grained siltstones and shales occupy the lower slopes. Where nearly flat-lying strata of slightly different resistance are exposed to weathering processes and erosion, weaker shaly rocks below the sandstone give way while the sandstone maintains the heights of hills. Forest vegetation types are similarly distributed with various trees and other plants preferring soils developed on certain rocks.

Figure 249. One of many vistas of rolling hills dominating Brown County State Park. Hills on the horizon are capped by more resistant sandstones.

Figure 250. Ravines carve smooth slopes in Brown County State Park.

Figure 251. Relative resistance to weathering separates sandstone/siltstone from silty shale (Brown County State Park)

Figure 252. Ogle Lake. Siltstones/sandstones/silty shales are exposed on Trail 7 leading to the lake. (Brown County State Park)

More detailed geologic setting of Brown County State Park

Seas that covered Indiana in early and middle Paleozoic time took a pause at the end of Devonian and start of Mississippian. A dead zone in the sea stretched across many states. Oxygen levels in this dead sea were very low. Few organisms could live in an environment similar to the Black Sea today. The New Albany black shale is one of these pauses. Instead of limestone, sandstone and ordinary shale, pyrite (fool's gold) and organic matter accumulated with clay in these stagnant waters. Once black shale was buried, organic materials turned into natural gas and liquid petroleum. These gases and liquids leaked out of the shale and accumulated in permeable sandstones and limestones waiting for a petroleum geologist's drill bit to liberate them. Your auto and home are likely powered by this fuel.

The seas returned to normal seawater conditions in early Mississippian. Fossiliferous Rockford Limestone was deposited. Uplift in the Appalachian Mountain system shifted the shoreline further west, replacing the limy sea with silt, clay, and sand washed off highlands into a vast delta (Borden Delta). The delta consisted of lobe-like masses as the river system grew out into the sea like the Mississippi delta does today. The New Providence Shale was deposited in deeper water (*prodelta* environment at the outer edge of a lobe) in front of the delta system. This shale was mainly clay. The Spickert Knob Formation accumulated on the sloping front of the delta (*delta front*). This layer was a mixture of silt and clay with some sand. The Edwardsville Formation was deposited nearest the coastline (*delta plain*). This stratum is mainly silt and sand with some clay. The total delta deposits are 500 to 800 feet thick. As the delta extended into the sea, these three formations were stacked over each other. Fine-grained prodelta sediments were buried by larger-grained delta front deposits. The delta front was covered by delta plain coarse-grained sediments. This explains the increase in grain size of the sediments from New Providence to Spickert Knob to Edwardsville Formation as time passed (Figure 253).

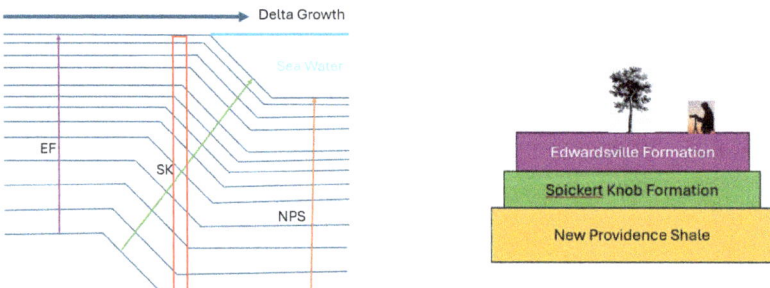

Figure 253. Cross-section illustrates how delta layers grow with time. EF=Edwardsville, SK=Spickert Knob, NPS=New Providence. Red box is hillside outcrop where EF, SK, and NPS overlie each other as in symbolic Brown County Hill (right). Sketches are vertically exaggerated.

Stream erosion of these stacked formations yields the Brown County State Park landscape. Rounded and steep-sided hills are held up by the Edwardsville sandstone. Weaker fine-grained strata are protected by the resistant caprock (Figure 253). Borden Delta's resistant rocks are expressed as the Knobstone Escarpment. Weed Patch Hill is the highest point in southern Indiana and the fourth highest point in Indiana at 1,058 feet. The three tallest hills are all within the Wisconsinan glaciated area.

The park was never touched directly by pre-Illinoian, Illinoian, or Wisconsinan glaciation. The Knobstone Escarpment served as a buttress to the ice. There is evidence that Illinoian ice crept onto the eastern edge of the escarpment but with little effect. Salt Creek did receive Illinoian glacial meltwater sediments, partly filling the valley. Most of these have been eroded in post-glacial times. Downcutting produced terraces which resulted from the meandering creek swinging from one side of the valley to the other. Terrace deposits with flat tops are the residuals from this stream erosion and downcutting.

Example nature preserves in Norman Upland

Scout Ridge Nature Preserve lies at the edge of Illinoian glacial evidence and includes glacial erratics brought in from Canada. Mississippian limestone and shale are the underlying bedrocks. The glacier extended into valleys without leveling the hills. Glacial boulders are present in the streambed. This preserve is in the fuzzy transition zone between Martinsville Hills and Norman Upland.

Hemlock Bluff Nature Preserve displays a steep bluff where Guthrie Creek exposes Mississippian shale.

Griffy Lake Nature Preserve near Bloomington contains **Cascade Park Trail** which leads to a waterfall developed on Mississippian siltstones. The waterfall is on a small tributary to Griffy Creek which has been dammed to form Griffy Lake.

4g. Scottsburg Lowland

Pre-Wisconsinan glaciers covered the Scottsburg Lowland. The northeastern and northern boundaries are defined by the southern edge of Wisconsinan till deposits. The sharp western boundary is located at the base of the Knobstone Escarpment (Figure 254). The southern boundary is at the drainage divide between the Muscatatuck River and Silver Creek. The eastern boundary, which is not sharp, divides the low flat Scottsburg Lowland from the flat slightly sloping limestone uplands of the Muscatatuck Plateau.

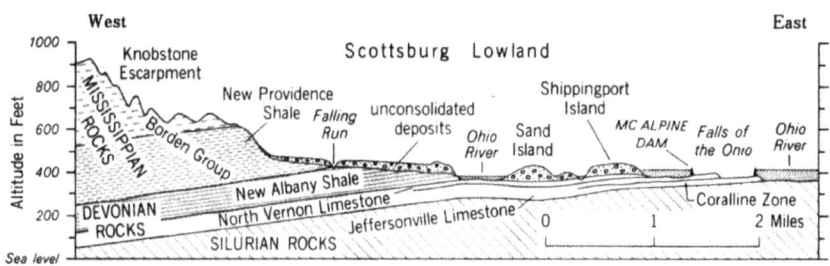

Figure 254. Profile illustrating the contrast between the Norman Upland (Knobstone Escarpment) and the Scottsburg Lowland. Powell, R., 1970, Geology of the Falls of the Ohio River: Indiana Geological Survey Circular 10, p. 19, fig. 8. doi: 10.5967/e9s7-c442. © Indiana Geological and Water Survey, Indiana University, Bloomington.

The East Fork of White River is the main stream in the Scottsburg Lowland. This river maintains a broad floodplain with active meanders and low terraces underlain by Wisconsinan meltwater sand and gravel (Figure 255). Sand dunes formed at the edges of terraces, notably in the east.

Geology of Indiana State Parks

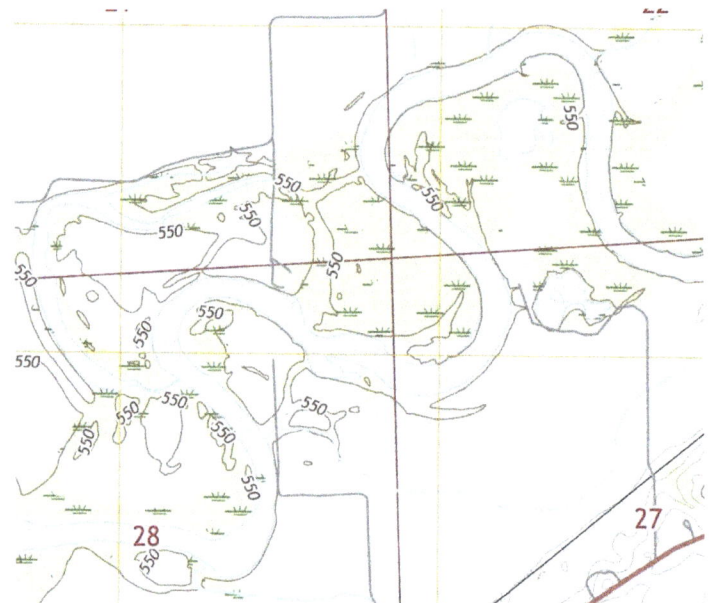

Figure 255. Flat floodplain with cutoff meanders and swamps of the East Fork of the White River. (U.S. Geological Survey Seymour Quadrangle, (2022))

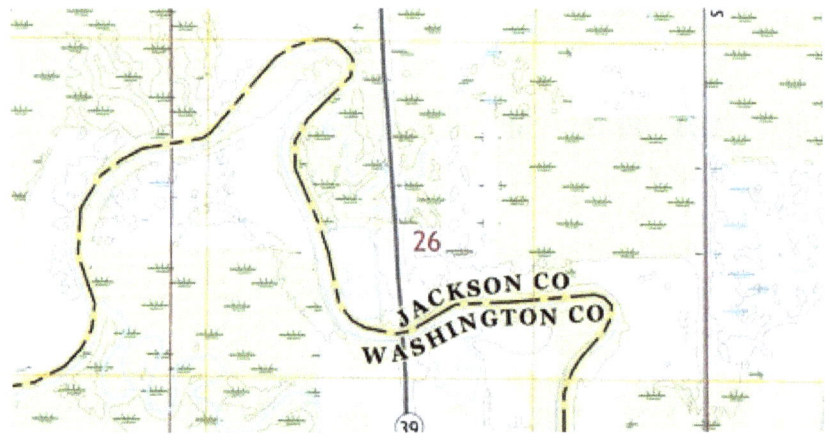

Figure 256. Flat, swampy floodplain of the Muscatatuck River. (U.S. Geological Survey Tampico Quadrangle, (2016))

The Muscatatuck River crossing the southern part of the lowland had no benefit from Wisconsinan meltwater. As a result, this stream carried no sand to speak of. The floodplain and terraces (Figure 256) consist of silt and clay, partly deposited in a lake.

Extensive lowland flats lie between and near major streams. Low hills capped by pre-Wisconsinan till are along the edges of the section. Bedrock is weakly resistant shale of Mississippian and Devonian age, producing few outcrops.

A unique area with extensive bedrock outcrops is the Brownstown Hills, just southeast of Brownstown. This small, rugged area is underlain by Borden Group siltstone/sandstone more resistant to erosion than the lowland's shale (Figure 257). These hills resemble isolated hills (The Knobs) to the north and are remnants of the Norman Uplands.

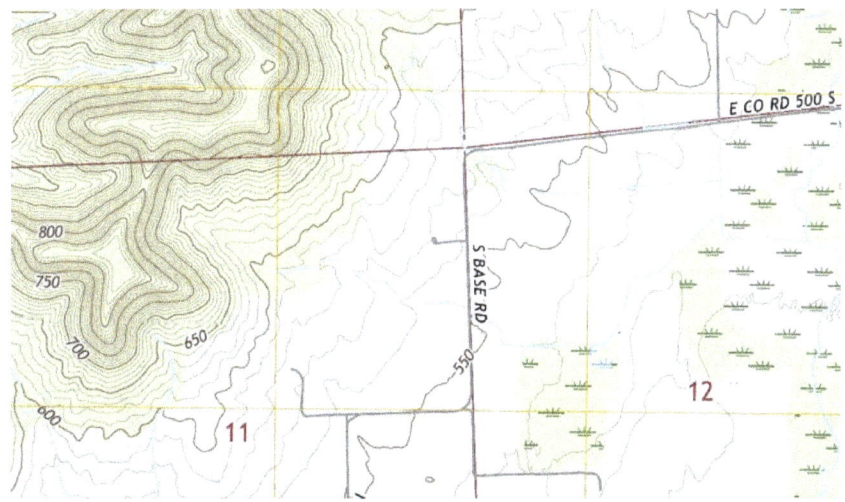

Figure 257. Brownstown Hills, an erosional remnant of the Borden siltstone/sandstone strata surrounded by Scottsburg Lowland shale. (U.S. Geological Survey Vallionia Quadrangle, (2016))

The Scottsburg Lowland and Norman Upland are textbook examples of how bedrock strongly influences landscape. The upland with its younger resistant siltstone/sandstone sports the most rugged Indiana topography. The lowland with its older weak shale struggles to attain much topographic relief.

The Scottsburg Lowland is bounded on the north by the Wisconsinan glacial boundary (New Castle Till Plains and Drainageways), on the west by the Norman Upland (Knobstone Escarpment), on the east by the Muscatatuck Plateau, and on the southeast by the Charlestown Hills (Figure 2).

There are no state parks in Scottsburg Lowland

Example nature preserve in Scottsburg Lowland

Thomastown Bottoms Nature Preserve is an extensive hardwood bottomland forest on the Muscatatuck River floodplain.

Geology of Indiana State Parks

4h. Charlestown Hills

The Charlestown Hills is south of the Scottsburg Lowland. It is separate from the Scottsburg Lowland by geology and landscape. Its western boundary is the base of the Knobstone Escarpment (Figure 258). The northern boundary is the southern limit of some pre-Wisconsinan glacial deposits that affected the Scottsburg Lowland. On the northeast, the Charlestown Hills is bounded by the Muscatatuck Plateau, where limestone uplands rise above the low shale hills.

Figure 258. Knobstone Escarpment separates the Norman Upland on the west from the Charlestown Hills on the east. U.S. Geological Survey New Albany Quadrangle, 2016)

Like the Scottsburg Lowland, the Charlestown Hills section is mainly underlain by soft Devonian and Mississippian shales. Outcrops are common since till cover is slight. Silurian and Ordovician strata are locally exposed. This entire section has been covered by pre-Wisconsinan glaciers, but little remains of the till. Glacial outwash terraces block Silver Creek, causing lake silt and clay deposition in the valley.

The Charlestown Hills is bounded on the north by the Scottsburg Lowland, on the east by the Muscatatuck Plateau, on the south by the Ohio River, and on the west by the Norman Upland (Knobstone Escarpment) (Figure 2).

Charlestown State Park

Charlestown State Park lies on the edge of the Ohio River valley. This 5,118-acre park was once part of the 15,000-acre Indiana Army Ammunitions Plant that operated from 1940-1945. The area began as a recreational site in 1995 and eventually became the present public state park in 1996. It has hills and ravines with 200 feet of local relief. Seventy-two bird species have been observed in the park. The park is the third largest in Indiana. Views of the Ohio River and Fourteenmile Creek are scenic. Fourteenmile Creek is touted as the oldest unglaciated stream valley in Indiana (Figure 259). Locks and dams on the Ohio River back up water to convert Fourteenmile Creek into a lake.

Figure 259. Fourteenmile Creek at IN 62 bridge, before the stream enters Charlestown State Park. Point bars on the inside of two meanders and a cutbank are visible on the outside of the nearest meander.

The following discussion is based primarily on Conkin et al. (1998). This publication describes features visible shortly after the park opened. Since then, vegetation has covered many outcrops that were visible from the main road.

Thin Illinoian glacial tills are found outside and within the park. These consist of silt, sand, gravel and erratic boulders. Devonian strata tend to occupy the hilltops. Silurian rocks are found at lower elevations and Ordovician rocks are at still lower elevations on the hills. For example, Trail 1 begins with Devonian limestone and moves downhill toward Fourteenmile Creek revealing Silurian limestone. At the lowest elevations on the trail are Ordovician dolostones and shales. Other trails reveal similar exposures depending on the elevation. Trail 6 is noted for its waterfall.

Devonian limestones are pockmarked by karst sinkhole topography over much of the surface. Silurian outcrops of limestone, dolomitic limestone, chert and shale are concentrated along the Ohio River bluffs (Figures 260 and 261).

Ordovician limestone, dolomitic limestone, and shale are most common along Fourteenmile Creek. **Fourteenmile Creek Nature Preserve** lies within the park and preserves a rugged, incised landscape with limestone cliffs, deep ravines, caves, sinkholes, and small waterfalls.

Trail 3 leads to the Portersville Bridge over Fourteenmile Creek that connects to the 1920s Rose Island amusement park. A 1937 flood damaged the park beyond repair. Rose Island is on a peninsula, not an island.

Geology of Indiana State Parks

Figure 260. Bluffs of the Ohio River display primarily Devonian and Silurian limestone strata. Ordovician rocks are exposed in the lower reaches of Fourteenmile Creek and where the creek enters the Ohio River. (Charlestown State Park)

Figure 261. Bluffs in Kentucky across from Charlestown State Park serve as a backdrop for a river barge on the Ohio River.

Falls of the Ohio State Park

We are indebted to Katherine Bulinski, Bellarmine University, and Alan Goldstein, Falls of the Ohio State Park, for identification of fossils and corrections to this section.

This well-known park is the smallest of Indiana's parks at 165 acres. Origin Park is a closely related urban park project that is related to Falls of the Ohio State Park. The state park's location on the Ohio River preserves a large exposure of Middle Devonian Jeffersonville Limestone that is rich in fossils. The "Falls" is a 26-foot drop over 2.5 miles that is a long series of rapids created by the Ohio River running over resistant limestone layers (Figure 262).

Figure 262. Falls of the Ohio. Resistant Middle Devonian Jeffersonville Limestone made rapids in the river that disrupted early river traffic. This was circumvented by a canal to allow ships to pass downriver. (Falls of the Ohio State Park)

The "Falls" is the only place on the 981-mile length of the Ohio River where dangerous rapids impeded commercial boats. Early travelers faced rapids or low water rock barricades. Canal construction around the Falls solved the problem.

Fossils at this site have been well studied and arranged into six *biozones* (stratigraphic units defined by a fossil or collection of fossils). Corals (Figures 263-268), brachiopods, stromatoporoids, trilobites, bryozoans, echinoderms, and gastropods (snails) (Figure 269) are the fossil types found there. They represent organisms living in Middle Devonian seas. Occasional storms stirred the sediment and animals, breaking shells and exoskeletons into small pieces (Figure 270). Burial by younger sediments compacted and sea water cemented the sediment into limestone. Much later, groundwaters dissolved small caves in the limestone (Figure 271). Some 600 fossil species are known at this site with many being first described in collections from the exposure. No collecting allowed.

Rugose corals may be solitary "horn corals" (Figure 263) or colonial types (Figure 264). As a colony, rugose corals lived as separate polyps ("pipe organ coral") or shared walls-similar to the famous Michigan Petoskey Stones. Both solitary and colonial rugose corals have *septa* (vertical radiating struts to support the internal stomach folds) (Figure 263). Solitary types are the most common rugose type in the park.

Tabulate corals are colonial with adjacent polyps sharing a wall with neighbors on five or six sides ("honeycomb" corals). Individual polyps typically lack septa or are not very noticeable (Figures 265-268). Each individual of the colony displays horizontal *tabulae* (surfaces supporting soft parts). As polyps grew, new tabulae and walls were precipitated to match individuals next to them. Coral larvae swam to a location and settled on the seafloor. Colonial varieties budded to clone identical neighbors.

Geology of Indiana State Parks

Figure 263. Middle Devonian rugose solitary horn coral (possibly *Heliophyllum*) weathered out of limestone. Horizontal growth lines record the coral's vertical development. A partial circular cross-section (lower left) illustrates vertical septa, which supported coral's soft parts. (Falls of the Ohio State Park)

Figure 264. Middle Devonian *Eridophyllum seriale,* the largest colonial rugose coral in the upper fossil beds at the Falls of the Ohio State Park.

Figure 265. A tabulate colony (possibly *Favosites*) polished by the river. The colony was vertical in life. Some tabulate and rugose coral forms can be confusing but they have different internal structures. (Falls of the Ohio State Park)

Figure 266. Top view of a Middle Devonian tabulate colonial coral colony (*Pleurodictyum cylindricum*-commonly called "wasp's nest coral"). (Falls of the Ohio State Park)

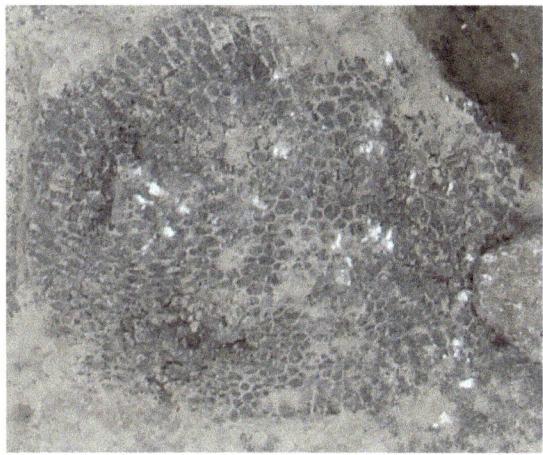

Figure 267. Top view of Middle Devonian tabulate colonial coral like Figure 268 (probably *Favosites)*. Note how polyps share walls with neighbors. (Falls of the Ohio State Park)

Figure 268. Longitudinal section of Middle Devonian tabulate colonial coral (probably *Favosites*). (Falls of Ohio State Park)

Geology of Indiana State Parks

Figure 269. Middle Devonian snail (*Turbinopsis shumar*-largest Devonian snail at site.). (Falls of the Ohio State Park)

Figure 270. Middle Devonian storm waves stirred shells, etc. breaking some into small pieces. (Falls of the Ohio State Park)

The best time to visit the park is August to November when the Ohio River is low. January to April is the worst time since this is high water time. The upper fossil beds are visible most of the year.

Illinoian glacial meltwater was mainly responsible for deepening the Ohio valley. Wisconsinan glacial meltwater spread alluvium over the area. Since that time, the Ohio River has been removing this sediment. As a result, these beautiful Devonian fossil beds have been exposed to the delight of paleontologists and the public alike.

Figure 271. Small cave dissolved in Middle Devonian limestone. (Falls of the Ohio State Park)

Be certain to see the interpretative center and its displays. Over 270 species of modern birds have been recorded in the park, as have over 100 species of fish in the river. The park was established in 1990, assuring that this exposure will not be vandalized and will be available for the public to view. Falls of the Ohio is a National Natural Landmark. Technically, this concentration of fossils is not a "reef" but a "biostrome" (Bulinski, personal communication, 2023). Biostromes are layered concentrations of marine fossils.

Example nature preserve in Charleston Hills

Nine Penny Branch Nature Preserve near Charlestown features deep ravines in Devonian limestone bedrock, small waterfalls, pools, riffles and slabs of limestone visible on a stream bed. There is ADA parking and a 1/3-mile ADA trail to an overlook. The trail continues for 0.4 mile to a small waterfall. The grade is ADA grade, but not with an ADA surface. The entire trail is a 1-mile round trip.

4i. Muscatatuck Plateau

Silurian and Devonian limestone and dolostone strata underlie Muscatatuck Plateau to yield a karst environment less impressive than the Mitchell Plateau. Karst is mainly found near deep incised streams. There is a thin cover of pre-Wisconsinan glacial deposits (Figure 272). Till is capped by silt that resembles loess.

Figure 272. Reddish soil developed on pre-Wisconsinan sediments of the Muscatatuck Plateau.

The northern boundary of the plateau is sharp where Wisconsinan glacial sediments make their appearance. The eastern boundary is where less steep limestone plateau rocks meet Dearborn Upland's narrow ridgetops and steep slopes. The western boundary is not sharp. Gently sloping plateau limestones meet flat Scottsburg Lowlands.

The Muscatatuck Plateau surface slopes westward at a rate less than the dip of the underlying Silurian and Devonian dolostone and limestone strata. This means that the land surface cuts off older and older rocks to the east. Many streams flow at a slight angle to the direction of slope of the plateau (Figure 273). This is due to pre-Wisconsinan glaciation.

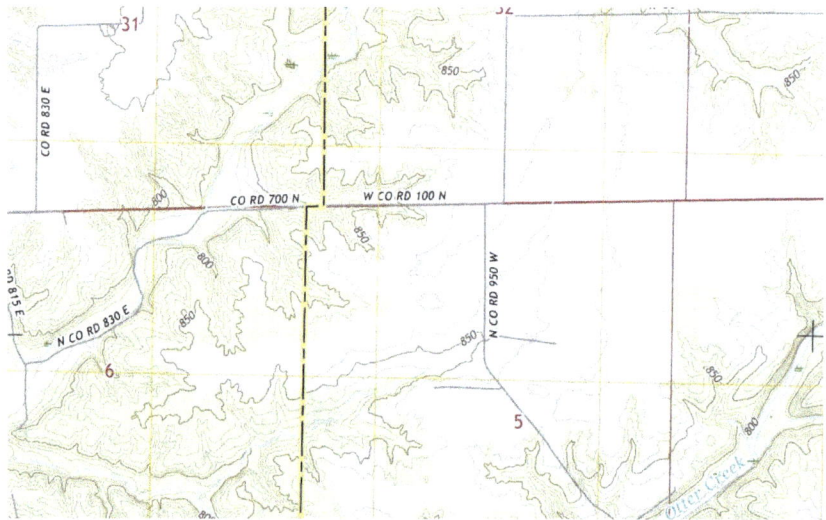

Figure 273. Streams on Muscatatuck Plateau flow at an angle to the westward dip of strata due to the influence of pre-Wisconsinan glaciation. Relatively gentle slopes are characteristic of the Plateau, in contrast with the easterly Dearborn Upland (Figure 284). (U.S. Geological Survey Holton Quadrangle, (2016))

The Muscatatuck Plateau is bounded on the northwest by the New Castle Till Plains and Drainageways, on the east by the Dearborn Upland, and on the west by the Scottsburg Lowland and Charlestown Hills (Figure 2).

Versailles State Park

This park is the second largest in Indiana at 5,988 acres and was turned over to the state by the National Park Service in 1943. Karst landscape can be seen with sinkholes and exposures of Ordovician and Silurian limestone bedrock and seasonal waterfalls. Sinkholes are especially seen along Trail 1. Dry streambeds typical of karst regions indicate the loss of surface water to the underground drainage system of caves. The park is known for its mountain bike trails.

Meltwater from Illinoian glacial ice enlarged the valleys of Laughery Creek (Figures 274-276) and Fallen Timber Creek (Figure 277) as the deluge of water flowed toward the Ohio River.

These "entrenched" streams of rapidly flowing water cut their way through Illinoian till and loess to dig deep and expose Silurian limestone, Ordovician Whitewater limestone and limy shale, and Dillsboro limestones and shales.

Invertebrate fossils are abundant in exposed bedrock. No fossil collecting is allowed in Indiana state parks.

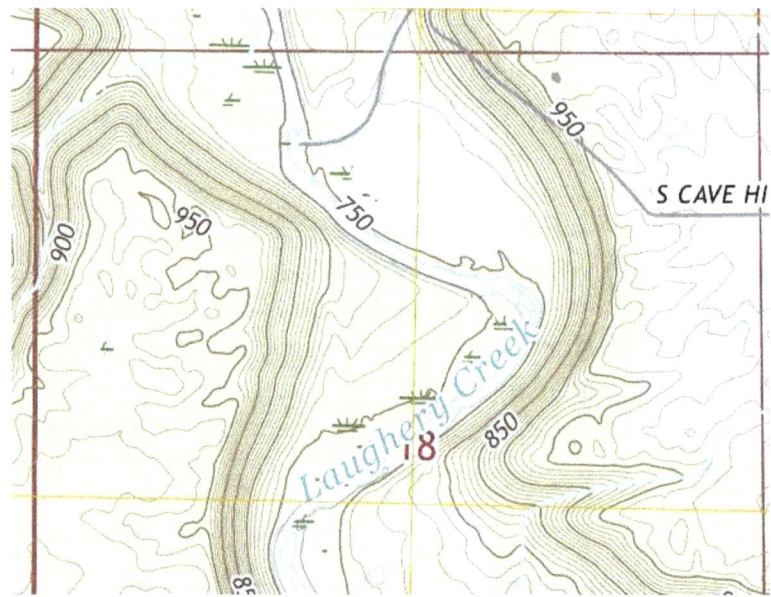

Figure 274. Deeply incised Laughery Creek. Illinoian glacial meltwaters enlarged the stream valley. Silurian and Ordovician strata are exposed in valley walls. Sinkholes dot the plateau. (U.S. Geological Survey Milan Quadrangle, (2016))

Figure 275. Covered bridge spans Laughery Creek. The valley was enlarged by Illinoian glacial meltwater, creating high cliffs. Silurian limestone cap the bluffs. Ordovician limestones and shales form the lower slopes. (Versailles State Park)

Geology of Indiana State Parks

The contact between Silurian and Ordovician rocks represents a change in the rock record due to erosion (unconformity). Shaly rocks of the Ordovician ended with this erosional break. This was due to glaciation in the southern hemisphere withdrawing huge amounts of seawater as continental glaciers formed. This lowered sea level, exposing the land to erosion by streams. Glaciers melted in Silurian time and the sea returned. This resulted in a significant change in the sediments being deposited as Silurian seas deepened and spread over most of North America. Less mud was deposited here, and abundant marine fossils resulted in limestone deposition.

Figure 276. Laughery Creek flows in its valley once buried by Illinoian glacial till. As the ice front melted back, rapidly flowing meltwater removed the till and deepened the valley. (Versailles State Park)

Fallen Timbers Nature Preserve lies within the park. Abundant fossiliferous rocks are exposed. Upland and riparian forests coat the landscape. Remember that no collecting of any kind is permitted in Indiana state parks.

Figure 277. Fallen Timbers Creek meanders at the base of its northern bluff (left). The creek was entrenched during Illinoian glacial meltwater flooding. (Versailles State Park)

Clifty Falls State Park

The third state park to enter the Indiana system (1920) was Clifty Falls. Its 1,416 acres bordering the Ohio River features several waterfalls. Sedimentary rocks in Clifty Falls span the end of the Ordovician Period and the beginning of the Silurian Period. This was a time of dramatic changes in sea level worldwide.

Ocean ridges separating continental plates rose during Ordovician time and displaced seawater, which resulted in some of the highest sea levels in the Earth's history. As a result, most of North America was submerged under a shallow sea. Ordovician sea levels generally stayed high with some fluctuations up and down.

Toward the end of the Ordovician Period, the Clifty Falls area received lots of mud and some limestone accumulation from the shallow sea. These sediments would eventually be compacted and cemented to form the Dillsboro Formation. This was followed by a clearing of the sea water as less mud was washed off the land. Abundant invertebrate animal shells accumulated. No collecting of fossils is permitted in Indiana state parks.

Ripple marks and shrinkage cracks in the limestone suggest the shallow sea was relatively clear of mud. Some of the limestone was eventually converted to dolostone by magnesium-rich water that changed calcite to dolomite minerals. These rocks formed the Saluda Member of the Whitewater Formation.

The last Ordovician stratum to be deposited over the Saluda was the Brassfield Limestone, found today in the southern part of the park. Later erosion removed the Brassfield from the northern section of the park. If you visit the waterfalls, the thick-bedded, resistant Saluda forms the caprock of major waterfalls, with weaker Dillsboro rocks worn back under this limestone ledge.

At the close of the Ordovician, a period of glaciation in the southern hemisphere removed an enormous mass of water from the oceans. This lowered sea level worldwide and initiated an unconformity (erosion surface). The Ordovician thus concluded with a massive drop in sea level of perhaps more than 500 feet. This drop exposed Ordovician sediments to surface streams and an eroded land surface formed.

The Silurian Period dawned with vast glacial sheets melting away. Meltwater poured into the ocean to raise sea level worldwide. Shallow seas again covered much of North America. In the northern Clifty Falls area, the first Silurian sediment to be deposited on the unconformity with Ordovician Saluda dolostone/limestone was the Osgood Member shale of the Salamonie Formation. This was followed by the Laurell Member limestone These strata are visible in the falls and cliff areas.

Now we skip forward hundreds of millions of years. Why? Because the sedimentary rocks deposited during the rest of the Silurian, Devonian, Mississippian, and Pennsylvanian were eventually eroded away as tectonic deformation in the Appalachians sent stresses that raised the Cincinnati Arch above sea level. Stream erosion removed this vast quantity of rocks. The erosion gap was perhaps 300 million years.

Not until the Pleistocene Epoch and its glaciation do we see much evidence of sedimentation in Indiana. Pre-Illinoian glaciers moved south and blocked north-flowing streams. Drainage shifted to the southeast as Illinoian glacial meltwater poured through the Clifty Falls area. The Ohio River developed with its deep and wide channel.

Like all tributaries entering a larger stream, Big Clifty Creek had to keep up with downcutting by the Ohio River. The creek incised into bedrock to create a three-mile-long canyon. This exposed, in succession downward, thin Illinoian glacial till, Silurian Laurel limestone, Osgood shale, and finally breached the erosion resistant Ordovician Saluda dolostone/limestone before slicing into the easily eroded Ordovician Dillsboro alternating shales and thin limestones.

Waterfalls in the park are formed by the resistant Ordovician Saluda dolostone/limestone overhanging the weakly resistant Ordovician Dillsboro shale/limestone. Examples of named waterfalls include Tunnel Falls (83 feet), Hoffman Falls (78 feet), Little Clifty Falls (60 feet), and Big Clifty Falls (60 feet). All these waterfalls are in **Clifty Canyon Nature Preserve** (Figures 278 and 282). How waterfalls form is discussed in "How Waterfalls Form" in the back of this book.

Trails offer excellent views of the gorge and rock outcrops. Trail 7 allows views of Big Clifty and Little Clifty Falls. Between these falls the trail leads to Cake Rock, a detached block of the Saluda which is an example of the interesting rock masses produced by weathering and erosion. Figures 278 and 279 illustrate the strata making up the falls. Figures 280-282 provide a sense of the immensity of rock removed by stream erosion during waterfall retreat.

IGWS (2018) notes that Big Clifty Canyon is receding at a rate of ¼ inch per year. Big Clifty Falls is estimated to have receded about two miles from the bluff of the Ohio River with the base of Big Clifty Creek entering the river about 226 feet lower than at the waterfall's inception (Indiana Department of Natural Resources quoted in Smath et al. (2017)). Remember to be careful around waterfalls due to their slippery and steep nature. Avoid climbing any waterfalls.

Figure 278. Big Clifty Falls. All falls in Clifty Falls State Park are developed on the resistant Saluda Member of the Whitewater Formation with underlying weak Dillsboro Formation strata that is easily eroded. Removal of the Dillsboro allows the overhanging layer to break off to cause the waterfall to recede.

Figure 279. Hoffman Falls. Saluda Member of the Whitewater Formation forms the resistant layer that is the lip of the waterfall overhanging the weaker Dillsboro Formation. Most Indiana waterfalls are seasonal and vary widely in their flow. Photo was taken during a dry season. (Clifty Falls State Park)

Figure 280. Waterfall retreat canyon downstream from Big Clifty Falls. As the waterfall retreats, erosion on the valley walls also wears away the sides. (Clifty Falls State Park)

Waterfall retreat involves downcutting of a tributary stream (Big Clifty Creek) to reach the elevation of the major stream (Ohio River). In addition, there is considerable erosion of the valley walls that involves weathering of the rocks and side-erosion by the tributary stream. When the tributary reaches a resistant layer (Saluda) then the waterfall can begin as erosion upstream is slower than downstream erosion where the Saluda has been worn away to expose the Dillsboro.

Figure 281. Waterfall retreat canyon of Big Clifty Creek. The entire space represented by the valley has been removed by retreat of the waterfalls and side-slope processes since the waterfalls started during incision of the Ohio River. (Clifty Falls State Park)

Geology of Indiana State Parks

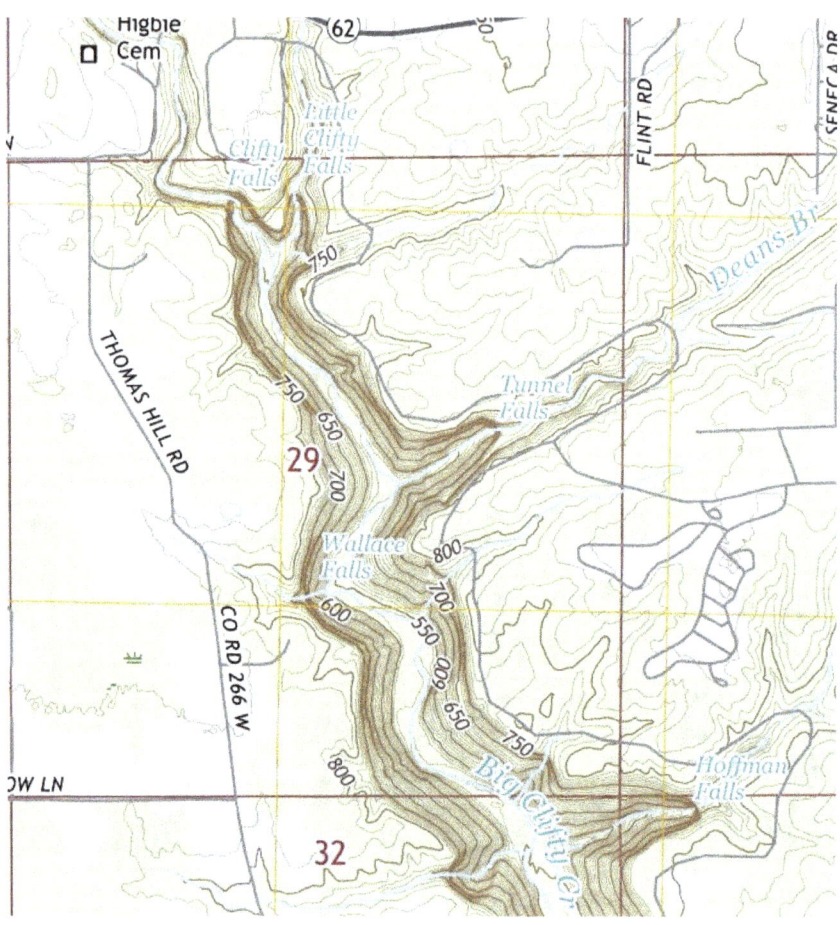

Figure 282. Falls of Clifty Falls State Park. Incision by Big Clifty Creek into the Plateau occurred due to downcutting of the Ohio River. (U.S. Geological Survey Clifty Falls Quadrangle, (2016))

Example nature preserves in Muscatatuck Plateau

Hanover College Gorge features several named waterfalls, e.g., Deadman Falls (30 feet), Horseshoe Falls, Crowe Falls, Butler Falls, Chain Mills Falls, and Fremont Falls (Figure 283).

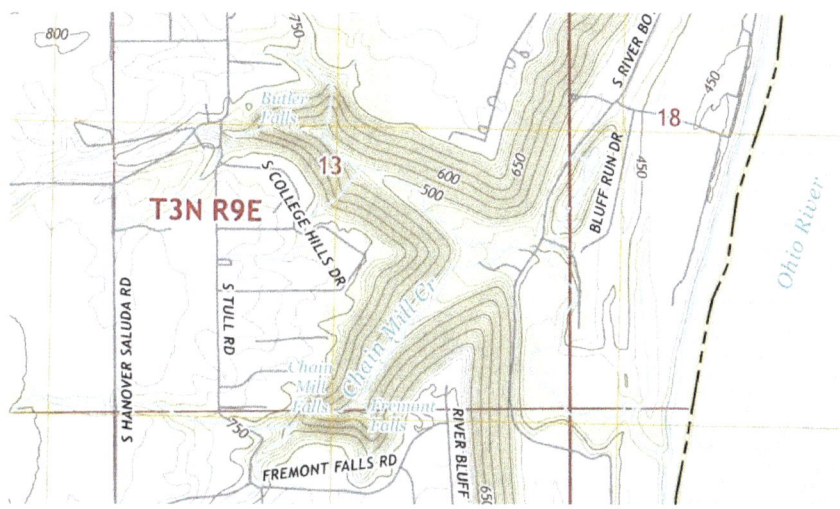

Figure 283. Waterfalls south of Hanover. Individual canyons are the result of waterfall retreat by streams as the Ohio River incised the landscape. (U.S. Geological Survey Madison West Quadrangle, (2016))

Calli Nature Preserve: North Vernon, Indiana, features cliffs and waterfalls (Rock Rest Falls) on Vernon Fork of Muscatatuck River.

Hardy Lake State Recreation Area is known for its limestone bluffs and caves.

Pennywort Cliffs Nature Preserve: Two springs join to feed a 30-foot waterfall before reaching Big Creek, Jefferson County's largest creek. There is no trail to the waterfall.

Muscatatuck Jennings County Park North Vernon, Indiana, displays 80-foot cliffs of Devonian strata. From top to bottom are North Vernon Limestone, Jeffersonville Limestone, and Geneva Member of the Jeffersonville at the base of the cliffs. Caves and waterfalls are present, including Muscatatuck Falls.

4j. Dearborn Upland

The Dearborn Upland is similar to the Norman and Crawford Uplands. Pre-Wisconsinan glaciers covered the Dearborn Upland and left sediments in many locations. This type of coverage differs from the other uplands of southern Indiana which received little or no pre-Wisconsinan glacial deposits. There are steep slopes, widespread *dendritic* ridgetop streams, and incision by a stream system. Local relief reaches 400 feet (Figure 284).

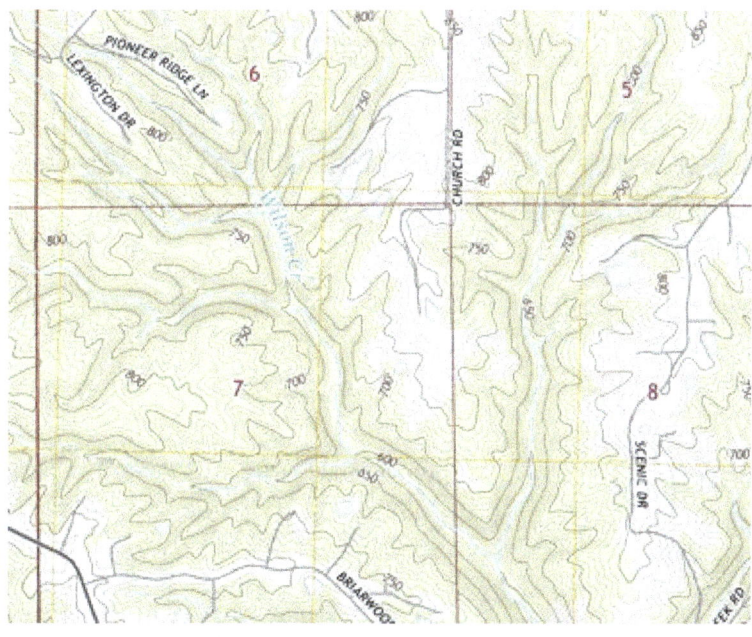

Figure 284. Typical Dearborn Upland topography. Contrast this rugged landscape of steep slopes and dendritic stream patterns with the gentle slopes of Muscatatuck Plateau in Figure 273. Both maps are about the same scale. (U.S. Geological Survey Aurora Quadrangle, (2016))

Bedrock is mainly shale with limestone beds. Drainage in southern Indiana prior to 700,000 years ago was to the north in the form of an Ancestral Kentucky River. Pre-Wisconsinan glaciation blocked this flow and forced water into what is now the Ohio River. Glacial meltwater deepened the channel and speeded up incision of the Dearborn Upland.

The Dearborn Upland is bounded on the north by New Castle Till Plains and Drainageways and on the west by Muscatatuck Plateau (Figure 2).

There are no state parks in Dearborn Upland.

Example nature preserve in Dearborn Upland

The Oxbow Nature Preserve is a meander cutoff of the Great Miami River where it empties into the Ohio. This floods annually and is a wildlife sanctuary (Figure 285).

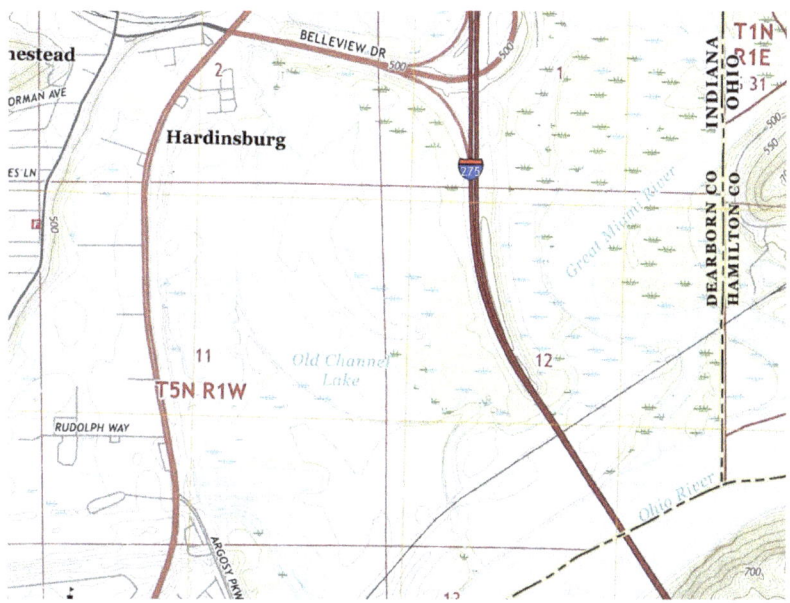

Figure 285. The Oxbow Nature Preserve resulted from a meander cutoff of the Great Miami River. (U.S. Geological Survey Lawrenceburg Quadrangle, (2016))

Conclusion

We hope you enjoy the wonderful state parks and nature preserves in Indiana. God made the landscape through the vastness of time and natural processes. Persons living in Indiana recognized the beauty and grandeur of features and organisms found throughout the state and were concerned that these special areas be preserved. The sensitivity of these individuals to their environment resulted in setting apart key sites to become parks and nature preserves. Their foresight was important to provide current and future generations with great places to relax and learn. Please continue this heritage by respecting natural features, plants, and wildlife in these beautiful locations so everyone can enjoy them! If we are careful, these amazing places can be available for future generations. We are all stewards of our planet.

Glossary of Terms

Algae: microscopic photosynthesizers that live in water.

Alluvial plain: a relatively flat surface formed by a stream depositing sediment.

Alluvium: sediment deposited by a stream.

Andesite: a lava rock intermediate in composition between rhyolite and basalt.

Anticline: upfold of strata.

Aragonite: mineral with the same composition as calcite; found in many shelled animals such as snails, clams, corals, etc.

Arch: a structural tectonic upwarp or underground tube or space open at both ends.

Artifact: object made or modified by humans.

Basalt: dark fine-grained igneous rock; common lava type.

Basement rocks: Precambrian rock encountered in a well drilled through younger rocks.

Basin: bedrock structure where all strata dip (tilt) toward the center of the feature.

Bedding plane: a surface separating two strata.

Bedrock: firm solid rock, underlies surface loose sediment and soil.

Biozone: a unique collection of fossils during a geologic time unit.

Blowout: a depression created by wind whipping out a low spot in sand dunes.

Bluff: steep surface at the edge of a valley.

Bog: lake being filled with vegetation and sediment.

Brachiopod: solitary invertebrate with two shells that are mirror images.

Braided river: a stream transporting mainly sand develops sand bars in the stream bed, splitting the river's flow.

Bryozoan: colonial invertebrates with individuals less than about one millimeter across that live on a stalk or fan.

Calcite: mineral making up many shells and all limestones ($CaCO_3$).

Cave: underground passage in soluble rock (limestone, dolostone, or gypsum).

Cave pearls: small, spherical grains of calcite deposited on sand in caves.

Cephalopods: predatory mollusks with tentacles; fossils with internal or external curved or straight shells.

Chert: chemical deposit of tiny quartz crystals.

Clams: solitary mollusk with two unlike shells.

Clay minerals: tiny platy minerals that form by weathering of other minerals.

Climate: average temperature and precipitation for an area.

Coal: buried swamp deposit of decomposed vegetable matter.

Column (Also see Geologic Column): speleothem where stalactite and stalagmite are joined.

Conchoidal fracture: smooth, curved surface where a mineral or rock breaks.

Conglomerate: gravel-size fragments cemented together to form a solid sedimentary rock.

Continental crust: thick outer layer of igneous, sedimentary, and metamorphic rocks making up the continents.

Coral: invertebrates with tentacle-bearing polyps; may secrete solitary horn-shapes or colonial reefs.

Crinoid: solitary marine invertebrate with food-gathering "arms", often attached to the seafloor; fossils display 5-sided symmetry.

Cross-bedding: angular lines in sedimentary rocks produced by

Cuesta: a hill or ridge of sedimentary rock with a steep front slope (escarpment) and a gentle back slope.

Currents depositing grains on a tilted surface.

Cut bank: eroded outside edge of a stream meander.

Cyclothem: rhythmic sequences of sedimentary rocks that repeat vertically due to changes in sea level.

Delta: a sedimentary environment accumulating where a stream enters a lake or ocean.

Delta front: the steep, seaward dropoff of a delta.

Delta plain: the flat upper surface of a delta.

Dendritic: tree-like map pattern of stream channels.

Deranged drainage: irregular stream patterns formed as a result of uneven deposition of glacial sediments.

Dike: a relative thin, tabular intrusion of igneous rock.

Dip: angle of tilt of layered rock measured from the horizontal.

Divide: divided terrain that separates drainage basins.

Dolomite: mineral making up dolostone; $CaMg(CO_3)_2$.

Dolostone: sedimentary rock of dolomite; altered limestone.

Dome: bedrock structure where all strata dip (tilt) away from the center of the feature.

Drapery: a narrow sheet speleothem growing from a cave roof.

End moraine: a moraine ridge that may not be recognized as recessional moraine or terminal moraine.

Epoch: a subdivision of a geologic period.

Era: a collection of geologic periods, e.g., Paleozoic.

Erosion: transportation of sediment by streams, wind, waves, glaciers, or gravity.

Erratic: a large glacially deposited boulder that does not match any local bedrock type.

Escarpment: steep edge of a plateau.

Esker: a sinuous ridge of gravel and sand that was a glacial tunnel filling.

Fault: fracture where rock blocks slide past each other.

Feldspar: common mineral in igneous or metamorphic rocks; weathers faster than quartz but can make up some sandstones and conglomerates; $KAlSi_3O_8$ (common feldspar in granite and rhyolite).

Floodplain: flat valley surface over which streams wander back and forth; floods during heavy rainfalls.

Fold: deformed rock layers, e.g., anticlines and synclines.

Forest: trees grow close enough that branches touch.

Formation: distinct layer of rock with characteristic features.

Fossil: preserved remains of an ancient life form.

Gabbro: a coarse-grained igneous rock equivalent of basalt.

Gastropod: a snail.

Geode: a hollow, elliptical rock precipitate, often with crystals inside.

Geologic Column: total collection of all rock units arranged in order by age.

Glacial: a specific Ice Age unit of time.

Glacier: thick mass of moving ice.

Glade: opening in a forest; often rocky with distinctive plants.

Gneiss: metamorphic rock with distinct layering of minerals.

Granite: massive, coarse-grained igneous rock mainly of feldspar and quartz.

Gravel: sedimentary grains larger than sand.

Ground moraine: an uneven deposit of till formed as a glacier receded.

Group: collection of similar formations.

Helectites: random crystal growth speleothems; often branched like tree roots.

Hill: a high area of any shape.

Hummocky topography: an uneven subdued landscape of rolling glacial landforms that is not as accentuated as distinct knobs and kettles, i.e., a subdued landscape akin to knob and kettle topography.

Ice Age: Pleistocene glaciation.

Igneous: rocks formed by cooling of molten masses.

Impact disturbance or structure: a shocked and distorted mass of rocks hit by an extraterrestrial mass.

Inner Core: the deepest part of the Earth, consisting of solid iron alloy; 3,200 to 4,000 miles.

Interglacial: a specific time unit between glacials.

Island arc: a string of volcanic islands near a plate boundary.

Joint: fracture where blocks of rock have not slid past each other; due to tension.

Kame: a gravel and sand hill formed by a stream flowing on top of a glacier that deposits sediment in a crevasse or low spot on the ice surface and is lowered onto the land as the ice melts away.

Kame and kettle topography: synonym for knob and kettle topography.

Karst: landscape marked by sinkholes and caves.

Karst window: a sinkhole resulting from collapse of a cave roof that exposes a cave stream.

Kettle: a depression on ground moraine surface formed by melting of an ice block broken from a receding glacier.

Kettle lake: a kettle with a lake.

Klint: Silurian fossil reef that is resistant to erosion and stands above the landscape as a hill.

Knickpoint: a sharp break in slope of a stream channel; often marked by a waterfall.

Knob: isolated hill of bedrock or glacial kame of gravel and sand.

Knob and kettle topography: landscape of kames and kettles.

Lamination: very thin layers in sediments or sedimentary rocks.

Land bridge: geologically temporary land mass connecting two continents, e.g., one that allowed Asian people to come to North America during the Pleistocene.

Landform: configuration of the land (e.g., hills, valleys, etc.).

Laurentide Ice Sheet: eastern glacial ice originating in Canada.

Lava: extrusion of molten liquid rock onto Earth's surface.

Liesegang rings: red/brown color bands resulting from chemical precipitation of iron in rock (usually sandstone) by groundwater.

Limestone: ground-up shells cemented by calcite to form a solid sedimentary rock.

Lobe: tongue-shaped mass of glacial ice.

Loess: silt-size, wind-blown sediment; usually blown off floodplains onto bluffs.

Longshore drift: movement of beach sediment by waves striking a beach at an angle.

Losing stream: synonym for sinking stream; part or all of the stream is captured by sinkholes or caves.

Magma: molten liquid rock under the Earth's surface.

Mammoth: extinct elephant-like, tusk-bearing mammal similar to a mastodon.

Mantle: rocky layer from the base of Earth's crust to 1,800 miles below the surface; thick layer over the outer core of the Earth; at least the upper layer is dynamic and moves along with the crust.

Marl: calcite deposited in a lake fed by springs rich in calcium carbonate; typically mixed with clay.

Mastodon: extinct elephant-like, tusk-bearing mammal similar to a mammoth.

Meander: curve in a stream channel.

Member: a stratum that is a subdivision of a formation.

Mesic: moderately wet ecological area.

Mesophytic forest: a highly diverse hardwood forest with rich soil.

Metamorphic rocks: have experienced high temperature and pressure by deep burial in the Earth which flattens and stretches minerals.

Mineral: naturally occurring inorganic solid with a definite chemical composition and repeating crystalline structure (e.g., calcite, quartz).

Mollusks: solitary invertebrates that live in water or damp environments; most have an external shell; snails, clams, cephalopods, etc.

Moraine: a glacial ridge of till.

Mound: rounded hill, often of Native American origin.

Mountain: tectonically produced uplift that may involve deformed rocks, igneous intrusions, and/or metamorphic rocks.

Mud: sediment of clay minerals and silt-size particles.

Native Americans: people living in the Americas since the last Pleistocene Glacial (about 11,700 years ago to the present; the Holocene); most likely descendants of Paleoindians.

Natural bridge (arch, tunnel); underground tube or space open at both ends.

Natural gas: flammable gas from decomposition of organic matter.

Ocean ridge: source of basalt in the ocean crust; once cooled, plates move away from the ridge; motions produce earthquakes.

Oceanic crust: relatively thin outer layer of Earth covered by ocean water and consisting of basalt and other dense rocks.

Outer Core: next layer out from inner core; made of liquid iron alloy; extends from 1,800 to 3,200 miles.

Outwash or outwash sediment: glacial meltwater sediment.

Oxbow lake: U-shaped lake formed where two stream meanders meet to cut off flow to a meander loop.

Paleoindians: an archeological term applied to humans living during the Pleistocene Ice Ages in North, Central, or South America; their descendants who live in the Holocene are called Native Americans.

Pangaea: a supercontinent that assembled during the Paleozoic Era and broke up after the Permian Period to separate into continents.

Peat: organic deposit in a standing water body.

Period: a unit of geologic time that is a subdivision of an Era.

Permeability: ability of a fluid to move easily through rock or sediment.

Petroleum: flammable liquid from decomposition of organic matter.

Physiographic province: large topographic unit with characteristic features.

Physiographic sections: subdivisions of physiographic provinces.

Physiography: the study of landforms.

Pinnacle: narrow landform resulting from erosion on two or more sides.

Pinnacle reef: an isolated Silurian fossil reef that is resistant to erosion and often forms a hill (klint) in glaciated areas.

Plate: rocky crust (oceanic or continental) and upper mantle of Earth.

Plate tectonics: the dynamic processes that involve movement of the ocean and continental plates to shift continents, generate earthquakes, form igneous rocks and metamorphic rocks, and produce mountains, volcanoes, and other major landforms (e.g., domes and basins).

Plateau: area of relatively level high ground; strata may be slightly tilted; often with deep valleys.

Point bar: deposition of alluvium on inside of a meander due to slowing of a stream to match water flow on outside of the meander.

Polyp: an individual coral in a colony or a separate organism.

Pothole: a cylindrical depression in bedrock of a streambed that is formed by gravel abrasion.

Prairie: temperate climate grassland.

Prodelta: the outer edge of a delta.

Pyrite (fool's gold): iron and sulfur mineral formed under very low oxygen conditions; often associated with coal; FeS_2.

Quarry: open pit mine for extracting bedrock.

Quartz: common mineral in igneous, metamorphic, and sedimentary rocks; very resistant to weathering; SiO_2.

Relief: maximum difference in elevation from one place to another; used to describe ruggedness of an area.

Rhyolite: fine-grained extrusive igneous rock; same composition as granite; requires high temperature to pour out as lava.

Ridge: long narrow hill or mountain.

Rimstone dam: speleothem holding a pool of water.

Rock: natural solid composed of minerals and formed by igneous, metamorphic, or sedimentary processes.

Rodinia: a Late Precambrian supercontinent.

Rugose: individual or colonial corals with septae; all were extinct by the end of the Permian Period.

Sand: smaller grain size than gravel but larger than silt.

Sand dune: a hill of sand formed by wind moving sand grains.

Sandstone: sedimentary rock with grains cemented by calcite or quartz; usually permeable and resists erosion.

Savanna: grassy plain with few trees.

Sediment: loose material composed of clay minerals, silt, sand, and/or gravel.

Sedimentary rocks: sediment that has been compacted and/or cemented by calcite or quartz, to make a solid rock.

Septa: a vertical strut that supports a rugose coral polyp.

Shale: sedimentary rock made of clay minerals and silt-size grains; highly compacted, often with laminations; not permeable.

Silt: tiny grains just barely visible to the unaided eye.

Siltstone: a sedimentary rock composed of cemented silt grains.

Sinkhole: surface depression in soluble rock (limestone, dolostone, or gypsum) by collapse of a cave roof or water dissolving rock along fractures.

Sinking stream: karst stream that disappears into a sinkhole or cave.

Snail: solitary mollusk, often with a spiraling shell.

Soda straw: tubular speleothem on cave ceilings; precursor of a stalactite.

Soil: loose surface material that coats most bedrock and sediments; formed by weathering due to activity of precipitation, plants, and animals; the medium in which most plants grow.

Source rock: organic-rich shale that provides petroleum and natural gas which accumulates in permeable rocks.

Speleothem: chemical cave deposit.

Spring: where groundwater visibly pours out onto Earth's surface.

Stalactite: speleothem that precipitates from saturated groundwater entering a cave ceiling; elongated conical shape.

Stalagmite: speleothem on a cave floor; deposited by groundwater dripping from a cave ceiling; usually below a stalactite.

Strata (stratum): layered sedimentary rocks.

Strip mine: surface mine for coal.

Stromatoporoids: a group of layered reef-building sponges that secreted a limy skeleton; most were extinct by the end of the Devonian Period.

Subduction: the process of an ocean plate sliding under another ocean plate or continental plate, generating volcanoes and earthquakes.

Syncline: downfold of strata.

Tabulae: table-like surface supports a tabulate colonial coral polyp.

Tabulate corals: colonial corals with individual polyps living in tubes that touch their neighboring tubes; all were extinct by the end of the Permian Period.

Talus: loose debris that accumulates at the base of a steep cliff by falling rocks.

Tectonic activity: earthquakes, volcanoes, and mountain building; usually due to movement of Earth's rocky plates.

Terminal moraine: a ridge of till which represents the farthest extent of a glacier before it melted away.

Terra rosa: reddish, highly leached soil derived from limestone or dolomite, typically in karst areas.

Terrace: a flat surface adjacent to a river which represents an earlier level of a floodplain remnant after stream incision through alluvium.

Till: loose, chaotic mixtures of gravel, sand, and mud deposited directly onto land as glacial ice melts.

Topography: "the lay of the land" or how land surface elevation changes from place to place.

Trilobite: solitary invertebrate with three distinct segments; they became extinct by the end of the Permian Period.

Tunnel (glacial): extended tubular opening at base of a glacier.

Unconformity: buried erosion surface.

Underfit stream: a small stream flowing in a large valley.

Valley: elongated low feature through which streams or glaciers move or have moved in the past.

Waterfall: stream water in free-fall, typically formed at a knickpoint.

Weathering: chemical, physical, and biological processes that alter rocks to produce clay minerals, silt, sand, gravel, and dissolved mineral matter.

Woodland: collection of trees without the limbs of each tree necessarily touching another tree.

References

Becker, L. (1974). *Silurian and Devonian rocks in Indiana southwest of the Cincinnati Arch.* Indiana Department of Natural Resources Geological Survey Bulletin 50, Indiana Department of Natural Resources Geological Survey.

Bieberman, D., & Esarey, R. (1946). *Stratigraphy of four deep wells in eastern Indiana, Report of Progress No. 1.* Indiana State Geological and Water Survey.

Boulding, J. (2022). *Pre-Illinoian Glacial Deposits in Indiana (3): Six Proposed Designations for Classification of Pre-Illinoian Deposits in the Central Midwest, BSW Working Paper 3, Version 1.0.*

Camp, M., & Richardson, G. (1999). *Roadside geology of Indiana.* Missoula, Montana: Mountain Press Publishing.

Conkin, J., Conkin, B., & Steinrock, L. (1998). *Geology of Charlestown State Park, University of Louisville Studies in Paleontology and Stratigraphy.* Kentucky: University of Louisville.

Counts, R., Monaghan, G., & Herrmann, E. (2014). Quaternary geology and geoarchaeology of the lower Ohio River Valley, southwestern Indiana. *56th Midwest Friends of the Pleistocene Field Conference*, (p. 78). Angel Mounds, Evansville, Indiana.

Curry, B., Hajic, E., Clark, J., Befus, K., Carrell, J., & Brown, S. (2014). The Kankakee Torrent and other large meltwater flooding events during the last deglaciation, Illinois, USA. *Quaternary Science Reviews, 90*, 22-36.

Curry, B., Lowell, T., Wang, H., & Anderson, A. (2018). Revised time-distance diagram for the Lake Michigan Lobe, Michigan Subepisode, Wisconsin Episode, Illinois, USA. In *Quaternary Glaciation of the Great Lakes Region, Process, Landforms, Sediments, and Chronology* (pp. 69-101). Kehew, A.E.; Curry, B.B.

Droste, J., & Keller, S. (1989). *Development of the Mississippian Pennsylvanian Unconformity in Indiana.* Indiana Department of Natural Resources.

Farlow, J., Steinmetz, J., & DeChurch, D. (2010). *Geology of the Late Neogene Pipe Creek Sinkhole (Grant County, Indiana).* Indiana Geological Survey.

Fleming, A., Farlow, J., Argast, A., Grammar, G., & and Prezbindowski, D. (2018). The Maumee Megaflood and the geomorphology, environmental geology, and Silurian–Holocene history of the upper Wabash Valley and vicinity, north-central Indiana. In L. Florea (Ed.), *GSA Field Guide, Ancient Oceans, Orogenic Uplifts, and Glacial Ice: Geologic Crossroads in America's Heartland.* doi:10.1130/2018.0051(12)

Frushour, S. (2012). *A guide to caves and karst of Indiana.* Bloomington and Indianapolis: Indiana University Press and Indiana Natural Science.

Gray, H. (1989). *Quaternary Geologic Map of Indiana*, Miscellaneous Map 49, Bloomington, Indiana: Indiana Geological and Water Survey, Indiana University.

Gray, H. (2000). *Physiographic Divisions of Indiana, Indiana Geological Survey Special Report 61.* Indiana Geological and Water Survey, Indiana University.

Gray, H. (2001). Map of Indiana Showing Physiographic Divisions, Miscellaneous Map 69. Bloomington, Indiana: Indiana Geological and Water Survey.

Gray, H., Ault, C., & Keller, S. (1987). Bedrock Geologic Map of Indiana. Bloomington, Indiana: Indiana Geological and Water Survey.

Higgs, S. (2016). *A guide to natural areas of southern Indiana.* Bloomington, Indiana: Indiana University Press.

Higgs, S. (2019). *A guide to natural areas of northern Indiana.* Bloomington, Indiana: Indiana University Press.

Hughs, P. (2022). Concept and global context of the glacial landforms from the Last Glacial Maximum, Ch.46. In *European Glacial Landscapes, Maximum Extent of Glaciations* (pp. 355-358).

Hunt, C. (1967). *Physiography of the United States.* W.H. Freeman and Company.

Indiana Geological and Water Survey (IGWS). (2018). *Indiana rocks! A guide to geologic sites in the Hoosier state.* Indiana Geological and Water Survey, Mountain Press.

Kehew, A., Beukem, S., Bird, B., & Kozlowski, A. (2005). Fast flow of the Lake Michigan Lobe: evidence from sediment-landform assemblages in southwestern Michigan, USA. *Quaternary Science Reviews, 24*(22), 2335-2353.

Kehew, A., Esch, J., Curry, B., Huot, S., Caron, J., Yellich, J., & Karki, S. (2017). Meltwater source for the Kankakee Torrent, Abstracts with Program, v.49. *Geological Society of America, 2017 annual meeting.* Seattle.

Klement, K. (1967). *Practical Classification of Reefs and Banks, Bioherms and Biostromes.* Am. Assoc. Petroleum Geologists.

Langenheim, R. J., Kiefer, J., Farvolden, R., & Garozzi, A. (1966). *Geologic Guidebook I, Guide to the geology of the Cagle's Mill Spillway, Turkey Run State Park and the Pennsylvanian sequence at Montezuma, Indiana.* Department of Geology, University of Illinois.

Loope, H., Antinao, J., Monaghan, G., Autio, R., Curry, B., Grimley, D., . . . Nash, T. (2018). *At the edge of the Laurentide Ice Sheet: Stratigraphy and chronology of glacial deposits in central Indiana, in Ancient Oceans, Orogenic Uplifts, and Glacial Ice: Geologic Crossroads in America's Heartland, GSA Field Guide 51.* (L. Florea, Ed.) doi:10.1130/2018.0051(11)

Malott, C. (1922). The physiography of Indiana. In W. Logan, E. Cummings, C. Malott, S. Visher, W. Tucker, & J. Reeves (Eds.), *Handbook of Indiana Geology* (pp. 59-256). Indiana Dept. of Conservation, Division of Geology.

Malott, C. (1932). *Lost River at Wesley Chapel Gulf, Proceedings of Indiana Academy of Science* (Vol. 41).

Malott, C. (1952). *The Swallow Holes of the Lost River, Orange County, Indiana, Proceedings of Indiana Academy of Science* (Vol. 61).

Potter, E., & Siever, R. (1956). Sources of Basal Pennsylvanian Sediments in the Eastern Interior Basin 1. Cross-Bedding. *Journal of Geology, 64*(3), 225-244.

Powell, R. (1970). *Geology of the Falls of the Ohio River.* Indiana Geological Survey.

Reams, M. (2013). *Geology of Illinois state parks.* Independently published. Retrieved from Kindle (Amazon)

Reams, M. (2022). *Ice Age giant mammals of the Midwest.* Independently published, Retrieved from Kindle (Amazon)

Reams, M., & Reams, C. (2021). *Waterfalls in Illinois.* Independently published. Retrieved from Kindle (Amazon)

Reams, M., & Reams, C. (2022). *Geology of Missouri state parks.* Independently published. Retrieved from Kindle (Amazon)

Rovey, C., & Balco, G. (2010). Periglacial climate at the 2.5 Ma onset of Northern Hemisphere glaciation inferred from the Whippoorwill Formation, northern Missouri, USA. *Quaternary Research*, 151-161.

Rovey, C., & Balco, G. (2011, December). Summary of Early and Middle Pleistocene Glaciations in Northern Missouri, USA. *Developments in Quaternary Science, 15*, 553-561. doi:10.1016/B978-0-444-53447-7.00043-X

Rudman, A., & Rupp, J. (1993). *Geophysical Properties of the Basement Rocks of Indiana, Geological Survey Special Report 55.* Department of Natural Resources, Indiana Geological and Water Survey.

Schneider, A. (1967). *The Tinley Moraine in Indiana* (Vol. 77). Indiana Academy of Science.

Shrock, R. (1929). The klintar of the Upper Wabash Valley in Northern Indiana. *Journal of Geology, 37*(1), 17-29.

Smath, R., Ettensohn, F., & Smath, M. (2017). *History and Geology of Madison, Indiana, and Clifty Falls State Park, Guidebook 1, Series XIII.* Lexington: Kentucky Geological Survey.

Strange, N. (2018). *The complete guide to Indiana state parks.* Bloomington, Indiana: Quarry Books, Indiana University Press.

Sturgeon, P., Loope, H., & Russel, K. (2017). Glacial Features of Indiana: Indiana Geological and Water Survey Digital Information Series, 14.

Thompson, T., Sowder, K., & Johnson, M. (2015). Generalized Stratigraphic Column of Indiana Bedrock, Poster 6A. Bloomington, Indiana: Indiana Geological Survey.

U.S. Geological Survey. (1992). South Bend East Quadrangle.

U.S. Geological Survey. (2016). Albany Quadrangle.

U.S. Geological Survey. (2016). Aurora Quadrangle.

U.S. Geological Survey. (2016). Bass Lake Quadrangle.

U.S. Geological Survey. (1973). Bonfield, IL, Quadrangle

U.S. Geological Survey. (2016). Bluffton Quadrangle.

U.S. Geological Survey. (2016). Brownstown Quadrangle.

U.S. Geological Survey. (2016). Clifty Falls Quadrangle.

U.S. Geological Survey. (2016). Corunna Quadrangle.

U.S. Geological Survey. (2016). Crandall Quadrangle.

U.S. Geological Survey. (2016). DePauw Quadrangle.

U.S. Geological Survey. (2016). Dune Acres Quadrangle.

U.S. Geological Survey. (2016). Enos Quadrangle.

U.S. Geological Survey. (2016). Georgia Quadrangle.

U.S. Geological Survey. (2016). Highland Quadrangle.

U.S. Geological Survey. (2016). Holton Quadrangle.

U.S. Geological Survey. (2016). Huntington Quadrangle.

U.S. Geological Survey. (2016). Iona Quadrangle.

U.S. Geological Survey. (2016). Jasonville Quadrangle.

U.S. Geological Survey. (2016). Lakeville Quadrangle.

U.S. Geological Survey. (2016). Largo Quadrangle.

U.S. Geological Survey. (2016). Lawrenceburg Quadrangle.

U.S. Geological Survey. (2016). Ligonier Quadrangle.

U.S. Geological Survey. (2016). Linton Quadrangle.

U.S. Geological Survey. (2016). Madison West Quadrangle.

U.S. Geological Survey. (2016). Martinsville Quadrangle.

U.S. Geological Survey. (2016). Milan Quadrangle.

U.S. Geological Survey. (2016). Mitchell Quadrangle.

U.S. Geological Survey. (2016). New Castle East Quadrangle.

U.S. Geological Survey. (2016). North Liberty and Lakeville Quadrangles.

U.S. Geological Survey. (2016). Oliver Lake Quadrangle.

U.S. Geological Survey. (2016). Orland Quadrangle.

U.S. Geological Survey. (2016). Ormas Quadrangle.

U.S. Geological Survey. (2016). Osceola Quadrangle.

U.S. Geological Survey. (2016). Plymouth Quadrangle.

U.S. Geological Survey. (2016). Solitude Quadrangle.

U.S. Geological Survey. (2016). Tampico Quadrangle.

U.S. Geological Survey. (2016). Vallionia Quadrangle.

U.S. Geological Survey. (2016). Williamsport Quadrangle.

U.S. Geological Survey. (2016). Woodburn North Quadrangle.

U.S. Geological Survey. (2017). Three Oaks MI/IN Quadrangle.

U.S. Geological Survey. (2019). Angola West quadrangle.

U.S. Geological Survey. (2019). *Area reports --Domestic: U.S. Geological Survey Minerals Yearbook* (Vol. II). Reston, Virginia: U.S. Geological Survey. doi:10.3133/mybvII

U.S. Geological Survey. (2022). Ege Quadrangle.

U.S. Geological Survey. (2022). Seymour Quadrangle.

Understanding Topographic Maps

The United States Geological Survey publishes maps illustrating Indiana topography. In this book, the variety used is the 7 ½ minute quadrangle, each of which encompasses 7 ½ minutes of longitude by 7 ½ minutes of latitude. On these maps, one inch on the map equals 2,000 feet on the land surface. This allows fairly small landforms to be shown. During cutting and pasting, this scale is not maintained for each illustration shown.

Green colors are forested areas, solid black are individual buildings, and brown curved contour lines are elevation lines above sea level. Below are examples of topographic features on these maps.

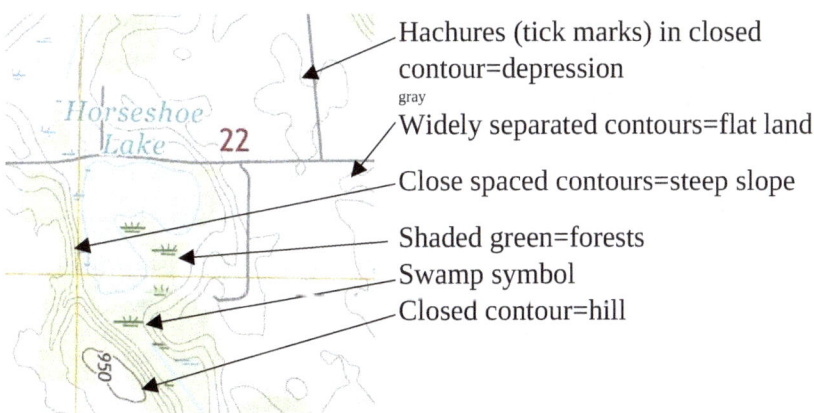

Hachures (tick marks) in closed contour=depression
gray
Widely separated contours=flat land
Close spaced contours=steep slope
Shaded green=forests
Swamp symbol
Closed contour=hill

U.S. Geological Survey Ormas Quadrangle, 2016.

How Waterfalls Form

There are many waterfalls in Indiana. A waterfall involves free-falling stream water. Waterfalls begin at knickpoints, geological events or rock characteristics that cause sharp increases in stream channel slopes.

Waterfalls form by a myriad of detailed situations. We will focus on the main types found in Indiana state parks and preserves. Most involve erosion-resistant limestone or dolostone (well-cemented sandstone or siltstone occur but are less common) that sit on top of weakly resistant shale or silty shales. Stream water pours over the resistant layer and goes into free-fall before striking the weak layer below. Weak rock wears away faster than resistant rock, causing the upper layer to extend as a ledge. Eventually, the lower layer cannot support the ledge. The ledge breaks off (often along a fracture) and topples into the stream bed below. The waterfall "retreats", producing a waterfall retreat canyon.

How do initial knickpoints form to begin this process? In Indiana, a major stream cuts deep into layered rocks. A tributary stream must also incise rocks to keep up with the larger stream. When the tributary reaches a resistant layer, this slows erosion of the tributary stream channel, perching the channel above the main stream. This forms a knickpoint as water falls over the resistant ledge to reach the main stream, eroding soft rock below. This initiates waterfall retreat as described above.

Exactly how undercutting occurs depends on characteristics of rocks. Examples are discussed in Reams and Reams (2021). Examples of possible processes in Indiana are illustrated here.

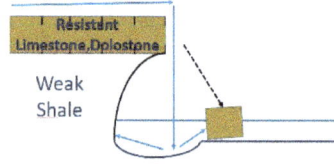

Waterfall Retreat: Unsupported stratum breaks off at fracture. Back splash of falling water erodes shale face.

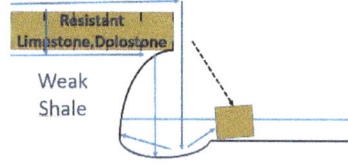

Waterfall Retreat: Unsupported stratum breaks off at fracture; groundwater flowing on top of shale erodes shale face, as well as falling water.

Index of Parks, Regions, and Preserves

Key references in the text for:
State Parks
Physiographic Regions
Nature Preserves

Always check online, by phone, or in person about safety issues, trails, weather, and availability of sites in advance of your visit. Never visit a site that is restricted or requires permission without first receiving clearance from appropriate authorities or agencies. Hunting may be permitted on some units. Be careful.

Ancient Pines Nature Area 96
Anderson Falls Nature Preserve 17
Angel Mounds State Historical Site 222
Auburn Morainal Complex 126
Barnes Nature Preserve 116
Beechwood Nature Preserve 139
Bender Nature Preserve 140
Big Spring Nature Preserve 263
Black Rock Nature Preserve 202
Blue Cast Spring Nature Preserve 146
Bluespring Caverns 246
Bluffs of Beaver Bend Nature Preserve 241
Bluffton Till Plain 147
Boonville Hills 218
Boot Lake Nature Preserve 119
Brookville Lake Dam spillway 173
Brown County State Park 269
Buzzard Roost Rec. Area, Hoosier National Forest, Tell City Unit 242
Cagles Mill Lake spillway 228
Cagles Mill Lake State Recreational Area and Natural Area 227
Calli Nature Preserve Cascade Park Trail 307
Cataract Falls State Recreation Area 225

Cave River Valley Natural Area 257
Cedar Bluffs Nature Preserve 258
Central Till Plain Region 147
Central Wabash Valley 182
Chain O' Lakes State Park 133
Charles McClue Reserve 139
Charlestown Hills 281
Charlestown State Park 283
Clifty Canyon Nature Preserve 301
Clifty Falls State Park 299
Clover Lick Barrens/Clover Lick S. A., Hoosier N. F., Tell City U. 242
Coastal Plain Ponds Nature Preserve 117
Conrad Savanna/Conrad Station Nature Preserves 116
Cowles Bog 95
Crawford Upland 232
Cressmoor Prairie Nature Preserve 103
Crooked Lake Nature Preserve 140
Davis-Purdue Agricultural Center Forest 155
Dearborn Upland 309
Donaldson Cave/Woods Nature Preserve 257
Dunes Nature Preserve 96
Eagle Marsh 155
Eagle Slough Natural Area 222
Fall Creek Gorge Nature Preserve 202
Fallen Timbers Nature Preserve 297
Falls of the Ohio State Park 286
Fawn River Nature Preserve 119
Fern Cliff Nature Preserve 199
Fisher Oak Savanna Nature Preserve 117
Flatwoods County Park 263
Flatwoods Nature Preserve 217
Fort Harrison State Park 178
Fourteenmile Creek Nature Preserve 284
Fox Island Nature Preserve 155
German Ridge Recreation Area, Hoosier Nat'l Forest, Tell City U. 242
Gibson Woods Nature Preserve 197
Granville Sand Barrens Nature Preserve 160

Greene-Sullivan State Forest 217
Green's Bluff Nature Preserve 242
Griffy Lake Nature Preserve 276
Hanging Rock Nature Preserve 153
Hanover College Gorge 307
Hardy Lake State Recreation Area 307
Harmonie State Park 214
Hathaway Preserve at Ross Run 153
Harrison-Crawford State Forest 237
Harrison Spring 242
Hayswood Nature Reserve 263
Hemlock Bluff Nature Preserve 276
Hemlock Cliffs Special Place, Hoosier National Forest, Patoka U. 241
Hoosier Prairie Nature Preserve 96
Hornbeam Nature Preserve 173
Hovey Lake Fish and Wildlife Area 217
Indiana Caverns 246
Indiana Dunes National Park 91
Indiana Dunes State Park 91
Iroquois Till Plain 156
Ivanhoe Dune and Swale Nature Preserve 97
Ivanhoe South Nature Preserve 97
John Merle Coulter Nature Preserve 99
Jug Rock Nature Preserve 243
Kankakee Drainageways 104
Kankakee Sands 114
Kokiwanee Nature Preserve 153
Lake Maxinkuckee 121
Lake Michigan Border 89
Leonard Spring Nature Park 258
Liebere State Recreational Area 225
Lincoln State Park 220
Little Calumet Headwaters Nature Preserve 103
Loblolly Marsh Nature Preserve 155
Lonidaw Nature Preserve 142
Lost River 258

Marengo Cave 243
Marion's Woods Nature Preserve 139
Martinsville Hills 223
Maumee Lake Plain 144
Maumee Lake Plain Region 144
McCormick's Creek State Park 250
Meno-aki Nature Preserve 155
Merry Lea Nature Preserve 140
Mitchell Karst Plains Nature Preserve 257
Mitchell Plateau 247
Moraine Nature Preserve 101
Mounds State Park 163
Mouth of the Blue River Nature Preserve 240
Muscatatuck Jennings County Park 308
Muscatatuck Plateau 293
New Castle Till Plains and Drainageways 161
Nine Penny Branch Nature Preserve 292
Norman Upland 265
Northern Moraine and Lake Region 83
O'Bannon Woods State Park 237
Olin Lake Nature Preserve 124
Orangeville Rise of the Lost River Nature Preserve 258
Ouabache State Park 149
Patoka Lake 246
Patoka Lake Hiking Area 241
Patoka River National Wildlife Refuge 222
Pine Hills Nature Preserve 197
Pinhook Bog 99
Plymouth Morainal Complex 120
Pokagon State Park 128
Porter West Preserve 242
Portland Arch Nature Preserve 199
Post-Oak-Cedar Nature Preserve 240
Potato Creek State Park 111
Potawatomi Nature Preserve 130
Prophetstown State Park 159

Raccoon State Recreation Area/Cecil H. HardenLake 199
Ritchey Woods Nature Preserve 181
Rocky Hollow Falls Canyon Nature Preserve 186
Ropchan Memorial Nature Preserve 123
Ropchan Wildlife Refuge/Ropchan Wetland Conservation Area 139
St. Joseph Drainageways 118
Salamonie River State Forest 153
Sand Hill Nature Preserve 109
Sarah Lincoln Woods Nature Preserve 220
Scottsburg Lowland 277
Scout Ridge Nature Preserve 276
Seidner Dune and Swale Nature Preserve 97
Seven Pillars Nature Preserve 150
Shades State Park 195
Shakamak Prairie Nature Preserve 213
Shakamak State Park 212
Southern Hills and Lowlands Region 206
Spicer Lake Nature Preserve 101
Spinn Prairie Nature Preserve 117
Spring Mill State Park 254
Springfield Fen Nature Preserve and Galena Marsh Wetland C. A. 103
Spurgeon Nature Preserve 124
Summit Lake State Park 167
Stout Woods Nature Preserve 181
Stoutsburg Savanna Nature Preserve 117
Tank Spring Nature Preserve 241
Teeple Glade Nature Preserve 263
Tefft Savanna Nature Preserve 117
The Cedars Preserve 258
Thomastown Bottoms Nature Preserve 280
Tippecanoe River Nature Preserve 109
Tippecanoe River State Park 108
Tipton Till Plain 158
Tolleston Strand Plain 97
Tom/Jane Dustin, R.C./R.C. Johnson, and Whitehurst N. Preserves 155
Twin Creek Valley Nature Preserve and Henderson Park 263
Thistlethwaite Falls 174

Trine State Recreation Area 128
Turkey Run State Park 185
Twin Swamps Nature Preserve 217
Valparaiso Morainal Complex 100
Vandolah Nature Preserve 155
Versailles State Park 295
Wabash Lowland 208
Wabash Lowlands Nature Preserve, Flatwoods Nature Preserve 217
Twin Swamps Nature Preserve 217
Warsaw Moraines and Drainageways 122
Wayne Woods 258
Weber Lake 221
Weiler-Leopold Nature Preserve 202
Wesley Chapel Gulf, Hoosier National Forest, Lost River Unit 263
Wesselman Woods Nature Preserve 222
White River State Park 177
Whitewater Memorial State Park 172
Whitewater Valley Gorge Park 174
Williamsport Falls 203
Wing Haven Nature Preserve 139
Wolf Cave Nature Preserve 253
Wyandotte Caves 240
Yellow Birch Ravine Nature Preserve 246

www.ingramcontent.com/pod-product-compliance
Lightning Source LLC
Chambersburg PA
CBHW052139220526
45471CB00004B/1446